Rainwater Harvesting for the 21st Century

Access to water in many parts of the world is increasingly challenging due to scarcity, quality issues and lack of access to adequate supply infrastructure. Currently, over 2 billion people around the world experience high water stress, and about 4 billion people experience severe water scarcity for at least one month on an annual basis. Rainwater harvesting (RWH) is increasingly seen as both an excellent alternative source of water and a valuable climate change adaptation measure. However, large-scale adoption remains challenging in many parts of the globe. This book, *Rainwater Harvesting for the 21st Century*, serves as a rigorous yet practical guide for a broad audience interested in the many opportunities that RWH systems can provide, including water and food security, flood management and climate change adaptation. It comprehensively covers the state of the art in RWH with practical examples of cutting-edge research and innovation in the design, operation and maintenance of RHW systems from both academics and practitioners. Highlights include:

- A comprehensive, transdisciplinary perspective of the latest advances in RWH techniques.
- Examples and case studies from around the world.

Rainwater Harvesting for the 21st Century

Edited by
Ilan Adler, Kemi Adeyeye, Aisha Bello-Dambatta
and Berill Takacs

CRC Press
Taylor & Francis Group
Boca Raton London New York

CRC Press is an imprint of the
Taylor & Francis Group, an **informa** business

Designed cover image: Shutterstock

First edition published 2025
by CRC Press
2385 NW Executive Center Drive, Suite 320, Boca Raton FL 33431

and by CRC Press
4 Park Square, Milton Park, Abingdon, Oxon, OX14 4RN

CRC Press is an imprint of Taylor & Francis Group, LLC

© 2025 selection and editorial matter, Ilan Adler, Kemi Adeyeye, Aisha Bello-Dambatta and Berill Takacs individual chapters, the contributors

Library of Congress Cataloging-in-Publication Data
Names: Bello-Dambatta, Aisha, editor. | Adeyeye, Kemi, 1979- editor. |
Adler, Ilan, editor. | Takacs, Berill, editor.
Title: Rainwater harvesting for the 21st century / edited by Aisha
Bello-Dambatta, Kemi Adeyeye, Ilan Adler and Berill Takacs.
Description: First edition. | Boca Raton, FL : CRC Press, 2025. |
Includes bibliographical references and index. |
Summary: "Rainwater Harvesting (RWH) is gaining much interest as both an alternative source of water and for its potential to be an important climate change adaptation measure, although large-scale adoption remains challenging in many parts of the world. Rainwater Harvesting for the 21st Century serves as a rigorous yet practical guide for a broad audience interested in the many opportunities that RWH systems can provide, including water security, food security, flood management, and climate change adaptation"– Provided by publisher.
Identifiers: LCCN 2024007936 | ISBN 9781032638089 (hardback) |
ISBN 9781032638096 (paperback) | ISBN 9781032638102 (ebook)
Subjects: LCSH: Water harvesting.
Classification: LCC TD418 .R367 2024 | DDC 628.1/42–dc23/eng/20240506
LC record available at https://lccn.loc.gov/2024007936

ISBN: 978-1-032-63808-9 (hbk)
ISBN: 978-1-032-63809-6 (pbk)
ISBN: 978-1-032-63810-2 (ebk)

DOI: 10.1201/9781032638102

Typeset in Times
by codeMantra

Contents

PART 2   Rainwater Harvesting in Practice

PART 3   Blueprint for the Future

Foreword

John Gould, Ilan Adler, Harry Chaplin

Humankind has been harvesting rainwater in one way or another since the dawn of time. The San people of the Kalahari in Southern Africa famously store rainwater in ostrich eggshells and bury them for future use. This has helped them survive in this arid desert environment for tens of thousands of years. As the global population grows and the stresses on finite water resources increase, exacerbated by climate change, we need to reflect on how our ancestors survived during the many challenging periods they faced.

In their important book, *Dying Wisdom: The Rise, Fall and Potential of India's Traditional Water Harvesting Systems in 1997*, Anil Agarwal and Sunita Narain traced the history of rainwater harvesting in Southern Asia back over 3,000 years. This work reflects what has been a growing interest worldwide in reviving the ancient art of harvesting rainwater for both domestic use and agriculture, stretching back over 60 years. Since the 1960s, there has been work in Africa on rainwater collection and storage, initiated by Oxfam and the Intermediate Technology Group (now Practical Action). The large-scale Thai jar programme in Thailand and the Water Cellar (*Shuijiao*) 1-2-1 project started in Gansu Province, China, have provided reliable, potable water supplies to several million poor rural families.

While the success of these projects has helped to spur interest in rainwater harvesting around the world, the lion's share of funding for water resource development has always gone to large-scale centralised dam construction, reticulated groundwater schemes and desalination projects. Small-scale decentralised water projects, whether wells, springs, gravity-fed supplies or rainwater harvesting, despite having generally been efficient, cost-effective and sustainable, have remained on the margins, offering little in the means of payback to the market economy.

This book follows a long list of comprehensive publications on rainwater harvesting, including, among many other excellent texts: *Rainwater Harvesting: The collection of rainfall and runoff in rural areas* by A. Pacey and A. Cullis in 1986; *Rainwater Catchment Systems for Domestic Supply: Design, Construction and Implementation* by J. Gould and E. Nissen-Petersen in 1999; *Making Water Everybody's Business: Practice and Policy of Water Harvesting,* A. Agarwal, S. Narain and I. Khurana (Editors) in 2001; *Roof Water Harvesting* by T.H. Thomas and D.B. Martinson in 2007 and *Every Last Drop: Rainwater Harvesting and Sustainable Technologies in rural China* by Z. Qiang, L. Yuanhong and J. Gould in 2012.

The publication of *Rainwater Harvesting for the 21st Century* is therefore a timely addition to these books, which have now also been supplemented by several excellent websites, such as the Centre for Science and Environment, India – *Rainwater Harvesting* site, and the University of Warwick, Development Technology Unit *Rainwater Harvesting – Domestic Roofwater Harvesting* site. These websites include much information on many aspects of rainwater harvesting systems, including design, water quality and even tools for calculating the optimum sizing and performance of different-sized catchments and tanks/reservoirs under different rainfall regimes.

As the world progresses through the third decade of the 21st century, almost all nations and communities are starting to realise that in the short term, before we can reduce greenhouse gas emissions to near zero and start removing these from the atmosphere at scale, we are going to need to adapt and become much more resilient to climatic extremes, whether droughts or floods. In this context, rainwater harvesting techniques offer multiple benefits and are clearly part of the solution to the challenges being faced, whether playing a role in flood control, firefighting, small-scale irrigation or domestic water supply.

In an environment increasingly filled with complex technologies, polarising politics and seemingly irreversible damage to the climate, rainwater harvesting is a beautifully simple, underutilised tool for collaboration, peacebuilding and an improved quality of life for all. Whether it's on a large or small scale, anyone can understand and follow the logic of collection and storage, making it a perfect example of a methodology apt for local design, adoption and replication.

This book is a wonderful amalgamation of stories and people from all walks of life, all over the world, who trust in rainwater harvesting as a solution to the multiple challenges we face. Its motivation began back in 2019 at a Symposium in Mexico City convened by the UK Royal Academy of Engineering (RAEng), where the editors met and discussed the relevance of combining so many interesting and global experiences in the field under the umbrella of a single publication. This was followed by the first International Conference on Rainwater Harvesting Systems in Madagascar, organised by Tatirano Social Enterprises, one of our leading contributors, with funding from both the RAEng and the UCL Global Engagement Office (University College London), to which the editors are greatly indebted.

This text is divided into three main sections: the first being about relevant research and innovation, covering a wide range of topics and perspectives, from life cycle analysis to remote monitoring technology and innovative design approaches. The second part focuses on Rainwater Harvesting in Practice, highlighting different studies related to the practical implementation of projects around the world. The third and final section, a 'Blueprint for the Future', looks at the bigger picture, including successful examples of financing mechanisms and the importance of strategic partnerships, showing that there is a hopeful future when it comes to rainwater harvesting and a great opportunity for further research and expansion.

Throughout each section, we have also threaded a number of case studies sent in by highly proactive NGOs and other organisations who are currently implementing projects in a variety of scenarios, hoping it will help readers have a glimpse into the wide variety of applications and environments where rainwater harvesting projects are possible, along with the substantial benefits they can bring to stakeholders.

About the Editors

Dr Ilan Adler is an associate professor in environmental engineering at University College London (UCL) as well as a Royal Academy of Engineering Industry Fellow. He is also the CEO and founder of EcoNomad Solutions, an enterprise dedicated to improving local conditions for farmers and rural communities through appropriate eco-technologies, such as waste-to-energy systems, among others. He founded the International Renewable Resources Institute (IRRI-Mexico) in 2002, an NGO initially dedicated to the dissemination of rainwater harvesting (RWH) in both rural and urban settings. He also sits on the Board of a number of organisations dedicated to the promotion and advancement of RWH, including Tatirano Social Enterprises (Madagascar) and Caminos de Agua (Mexico). He has conducted abundant research on the quality and modelling of RWH, with projects ranging from rural schools to high-rise buildings.

Dr Kemi Adeyeye is Associate Professor and Researcher at the Centre for Regenerative Design & Engineering for a Net Positive World (RENEW), the University of Bath. She specialises in multifactorial, multidisciplinary research using socio-technical methods. She has an established international profile in research at the crossroads of resource efficiency and resilience, investigating linkages between place, time, people, technology and policy. As PI or Co-I, she has undertaken many situated studies in developed and developing countries, in formal and informal contexts. She was a Leverhulme Research Fellow and led a DEFRA-funded collaborative network for over a decade. She is Co-Editor-in-Chief of the *International Journal of Architectural Engineering and Design Management*.

Dr Aisha Bello-Dambatta is an interdisciplinary researcher in water science and engineering. Her current research is broadly focused on the related areas of integrated water resources management, climate change impacts on land and water resources and water-energy-food-environment nexus. She has worked on several large-scale, international interdisciplinary collaborative research projects, including the EU-funded cross-border Dŵr Uisce water-energy nexus project, which worked to improve the energy performance of the water sectors in Ireland and Wales, as well as developing policy and best practice guidelines to facilitate the implementation of integrated low-carbon and smart energy solutions. She currently lectures in water sciences at University College London (UCL).

Dr Berill Takacs is an interdisciplinary researcher who is passionate about the environment and sustainable living and lifestyles. Her research interests lie in advancing interdisciplinary research in the fields of sustainable development, circular economy and sustainable consumption. With expertise in life cycle assessment (LCA), she has worked on projects modelling water-food-energy nexus and contributed to the development of tools assessing the environmental impacts and water footprint of food, meals and food loss and waste. Her current research focuses on environmental nutrition, with the aim of designing healthy and sustainable food systems that prioritise the well-being of humans, other species and the planet.

Contributors

Doaa Ismail Abuhamoor
Geographic Information System and Remote Sensing Department, Environment and Climate Change Directorate, National Agricultural Research Center (NARC), Amman, Jordan

Rania F. Aburamadan
Applied Science Private University, Architectural Engineering Department, Engineering Faculty

Kemi Adeyeye
Department of Architecture and Civil Engineering, University of Bath, UK

Ilan Adler
Department of Civil, Environmental & Geomatic Engineering, University College London (UCL), Gower Street, London WC1E 6BT, United Kingdom

Lubna Mustafa AlMahasneh
Geographic Information System and Remote Sensing Department, Environment and Climate Change Directorate, National Agricultural Research Center (NARC), Amman, Jordan

Juliana Farias Araujo
State University of Feira de Santana, Transnordestina Avenue, 44036-900, Department of Technology, Feira de Santana, Bahia, Brazil

Hamilton de Araújo Silva Neto
State University of Feira de Santana, Transnordestina Avenue, 44036900, Department of Technology, Feira de Santana, Bahia, Brazil

Admire T Baudi
Bolsan Trust, 4044 3rd Street, Gunhill, Dzivaresekwa, Harare, Zimbabwe

Aisha Bello-Dambatta
Department of Geography, University College, London, Gower Street, London WC1E 6BT

Harry Chaplin
Tatirano Social Enterprise, Fort-Dauphin, Madagascar

Eduardo Borges Cohim
State University of Feira de Santana, Transnordestina Avenue, 44036900, Department of Technology, Feira de Santana, Bahia, Brazil

K. Dimitriadis
Geoservice Ltd, 35 Lykaiou str., 11476, Athens, Greece

A. Eleftheriou
HYDRASPIS, 32 V. Tzelia, 43132, Karditsa, Greece

Abdoulaye Faty
Département de Geographie, Université Cheikh Anta DIOP de DAKAR, Senegal

Jeremy Gibberd
School of Construction Economics
and Management, University of the
Witwatersrand, Johannesburg, South
Africa and CSIR, Pretoria, South
Africa

John Gould
Department of Environmental
Management, Lincoln University,
New Zealand

Qais Hamarneh
Independent researcher

Jálvaro Santana da Hora
Federal Institute of Education, Science
and Technology of Bahia – Ilhéus
Campus, Jorge Amado Highway km
13, 45671-700, Technology Reference
Center, Ilhéus, Bahia, Brazil

Thiago Barbosa de Jesus
State University of Feira de Santana,
Transnordestina Avenue, 44036900,
Department of Technology, Feira de
Santana, Bahia, Brazil

Adriano Souza Leão
SENAI CIMATEC University
Center, Orlando Gomes Avenue
1845, 41650-010, Department of
Environment, Salvador, Bahia, Brazil

C. Makropoulos
Department of Water Resources and
Environmental Engineering, School
of Civil Engineering, National
Technical University of Athens, 5
Heroon Polytechneiou str., 15780,
Zografou, Athens, Greece

S. Malamis
Department of Water Resources and
Environmental Engineering, School
of Civil Engineering, National
Technical University of Athens, 5
Heroon Polytechneiou str., 15780,
Zografou, Athens, Greece

Lynn McGoff
Sustainable OneWorld Technologies,
83 Willingham Road, Over,
Cambridgeshire, CB24 5PF, UK

K. Monokrousou
Department of Water Resources and
Environmental Engineering, School
of Civil Engineering, National
Technical University of Athens, 5
Heroon Polytechneiou str., 15780,
Zografou, Athens, Greece

Michelle Morelos
Posgrado en Ciencias de la
Sostenibilidad, Universidad Nacional
Autónoma de México

David Morgan
University of Cambridge, UK

Sandile Mtetwa
University of Cambridge, Peterhouse,
Trumpington Street, Cambridge,
CB2 1RD, UK

John A J Mullett
Sustainable OneWorld Technologies,
83 Willingham Road, Over,
Cambridgeshire, CB24 5PF, UK

Francesca O'Hanlon
University of Cambridge, UK

Jorge A. Ortiz-Moreno
Institute of Development Studies,
University of Sussex

Laura Rodríguez-Bustos
Programa Universitario de Estudios
   Interdisicplinarios del Suelo,
   Universidad Nacional Autónoma de
   México

Emilio Rodríguez-Izquierdo
Ithaca Environmental

David Ruvubi
Department of Civil, Environmental &
   Geomatic Engineering, University
   College London (UCL), Gower
   Street, London WC1E 6BT, United
   Kingdom

Maria Auxiliadora Freitas dos Santos
Federal Institute Of Bahia, Science and
   Technology, Vicinal de Aparecida
   Street, 48700-000, Serrinha, Bahia,
   Bahia, Brazil

Edna dos Santos Almeida
SENAI CIMATEC University
   Center, Orlando Gomes Avenue
   1845, 41650-010, Department of
   Environment, Salvador, Bahia, Brazil

Isabel Schestak
School of Environmental and Natural
   Sciences, Bangor University,
   Bangor LL57 2UW, Gwynedd,
   United Kingdom

Jordan Silverman
Inclusive Energy, United Kingdom

Samuel Alex Sipert
State University of Feira de Santana,
   Transnordestina Avenue, 44036900,
   Department of Technology, Feira de
   Santana, Bahia, Brazil

Ana Caroline Bastos Lima de Souza
State University of Feira de Santana,
   Transnordestina Avenue, 44036-900,
   Department of Technology, Feira de
   Santana, Bahia, Brazil

Anderson de Souza Matos Gadéa
State University of Feira de Santana,
   Transnordestina Avenue, 44036-900,
   Department of Technology, Feira de
   Santana, Bahia, Brazil

M. Styllas
Geoservice Ltd, 35 Lykaiou str., 11476,
   Athens, Greece

Mohammad Talafha
Independent researcher

Beth Tellman
School of Geography, Development &
   Enviroment, University of Arizona

Dylan Terrell
Caminos de Agua, Mexico

I. Vasilakos
HYDRASPIS, 32 V. Tzelia, 43132,
   Karditsa, Greece

Karl Zimmermann
Environmental Engineering, University
   of British Columbia, Vancouver,
   Canada

Part 1

Research, Innovation,
State-of-the-Art

1 From Rooftops to Impermeable Pavements

The Lifecycle Assessment, Environmental Benefits and Trade-offs of Rainwater Harvesting

Isabel Schestak

1.1 INTRODUCTION

Since ancient times, rainwater harvesting (RWH) has been practised around the globe for drinking water supply, and in agriculture to secure food supplies (Mays, Antoniou, and Angelakis 2013; Ghimire and Johnston 2015). However, during the last decades, it has gained attention as an alternative or supplement to centralised municipal water supply in urban environments. It is seen as a means to reduce the growing pressure on urban water systems caused by rising population densities, diminishing surface or groundwater resources and climate change (Campisano et al. 2017; Teston et al. 2018; Sojka, Younos, and Crawford 2016). Energy intensity in water supply and related emissions of greenhouse gases (GHG) have been found to positively correlate with water stress due to the necessity of importing water from far basins or desalinating seawater (Lee et al. 2017). The increase in water stress due to climate change and urbanisation therefore calls for more energy-efficient supply strategies and the use of alternative local freshwater resources.

The potential benefits of harvesting rainwater are plentiful, including serving as a climate change adaptation measure. It can ensure sufficient water availability in otherwise water-scarce areas and alleviate or prevent overexploitation of aquifers used for groundwater supply (Valdez et al. 2016; de Sá Silva et al. 2022). When replacing potable mains water, the use of rainwater for non-drinking purposes like toilet flushing, laundry, or irrigation can reduce energy consumption for drinking water treatment and distribution by the water utility (Valdez et al. 2016; Ghimire et al. 2017). RWH is also a valuable tool in stormwater management (Scholz and Grabowiecki 2007; Lin et al. 2018; de Sá Silva et al. 2022). The retention of rainwater helps to prevent flood risks in cities exacerbated by the expansion of impermeable surfaces. It further reduces the amount of stormwater that needs to be treated

DOI: 10.1201/9781032638102-2

in a wastewater treatment plant (WWTP), thus reducing energy consumption for wastewater treatment. Reduced stormwater amounts prevent overflows of combined stormwater-wastewater sewer networks, which during peak flows can release harmful pollutants (Scholz and Grabowiecki 2007; Lin et al. 2018; de Sá Silva et al. 2022).

Rainwater harvesting, therefore, seems to be an obvious solution to water supply and stormwater management problems. However, installing a rainwater harvesting system (RWHS) can have detrimental consequences, too. Like conventional water supply and wastewater treatment systems, RWH comes with a burden through its infrastructure and construction activities, as well as its energy requirements during operation for pumping (if not entirely gravity-fed) or treatment (Ghimire et al. 2017; Angrill et al. 2012). Vieira et al. (2014) found the energy requirements for pumping in RWH could be as little as 0.04 kWh/m^3 or as much as 7.1 kWh/m^3. This is dependent on the design of the system, such as heights, pressure, piping, type of pumps, and water demand. The manufacture of the pump itself and pumping energy drove the environmental impacts of the RWHS. Other water supply systems can require energy in a similar range, from 0.0002 to 1.74 kWh/m^3 for surface water supplies and 2.4 to 8.5 kWh/m^3 for seawater desalination using reverse osmosis (Wakeel et al. 2016). Importing water resources over long distances, as practised in California, can increase energy intensity even further to 16.4 kWh/m^3 (Lee et al. 2017). A local supply of rainwater has the potential to reduce such energy consumption. The type of water source and its distribution is an influential factor for energy consumption in water supply and GHG emissions. Environmental assessments and considerate planning are therefore required to mitigate climate change contributions from the water sector and prevent burden shifting.

A suitable tool to evaluate the environmental benefits and trade-offs of water supply systems from a holistic perspective is life cycle assessment (LCA). LCA studies have been conducted for the water sector all over the globe, in particular for urban centralised municipal water supply and wastewater treatment systems (Loubet et al. 2014; Lee et al. 2017; Friedrich, Pillay, and Buckley 2009; Lundie, Peters, and Beavis 2004). The LCA methodology has been used to compare the environmental footprints of different water supply alternatives. Conventional municipal systems based on surface or groundwater sources have been compared to alternative supplies from desalination of sea- or brackish water, reclamation of wastewater, greywater reuse and also rainwater harvesting (Faragò et al. 2019; Godskesen et al. 2013; Amores et al. 2013; Ghimire et al. 2017; Leong et al. 2019). Due to the higher energy intensity of treating water to potable standards, wastewater reclamation and desalination might offer no significant environmental advantage or even increase impacts compared to municipal supply (Amores et al. 2013). In a study on a commercial building in Washington, DC, RWH was found to be the more favourable option in environmental terms for water supply for toilets and urinals compared to the municipal supply in most analyses (Ghimire et al. 2017).

Major environmental impacts on the global warming potential (GWP) of water supply systems were found to be caused by the use of electricity, especially for the distribution of drinking water in the network, water treatment and wastewater treatment (Zappone et al. 2014; Amores et al. 2013; Jeong, Minne, and Crittenden 2015; Xue et al. 2019). Electricity consumption, together with the use of water treatment chemicals, is also a major contributor to other impact categories such as eutrophication (Slagstad and Brattebø 2013). Infrastructure is another significant contributor to environmental impacts, not only on global warming but also on metal depletion (Xue et al. 2019), human toxicity or ozone depletion (Jeong, Minne, and Crittenden 2015).

This chapter introduces LCAs in rainwater harvesting. Its objective is to provide an overview of the range and variety of LCA studies concerning RWHS, rather than a systematic or full review of the literature on LCAs in rainwater harvesting. This chapter reviews relevant LCA studies on RWHS from the last decade, looking at potable and non-potable uses of rainwater, comparisons of RWHS with alternative water supply strategies, design analysis of the RWHS, different RWH scales, and the integration of RWH with other alternative water supply forms such as greywater recycling. The studies are discussed to highlight key factors influencing the environmental performance, from the physical design of the system to the behaviour of its users. Further, the pros and cons of RWH compared to conventional and other alternative water supply strategies are shown. The chapter finishes with a discussion and summary of the lessons learned for the sustainable design of RWHS.

1.2 METHOD

1.2.1 THE LIFE CYCLE ASSESSMENT METHOD

An LCA quantifies the environmental impacts caused by a product or service during its whole life cycle, from resource extraction to manufacturing, use and disposal or end-of-life (EoL) stage, based on emissions to the air, water and soil or due to resource consumption (Bjorn et al. 2018). Normally, a comprehensive LCA covers the whole life cycle, called a cradle-to-grave LCA. However, if the focus is on only a part of the life cycle or certain data is lacking, it might be cradle-to-gate (manufacturing of a product only) or cover the use phase only. A range of different environmental impact categories can be assessed, such as climate change (or global warming), aquatic or terrestrial eutrophication, acidification, human and ecotoxicity, particulate matter formation, ozone depletion, freshwater use, land occupation, or depletion of mineral, fossil and biotic resources. Through its holistic and comprehensive approach, trade-offs can be identified and burden shifting prevented, both between life cycle stages and between environmental impacts (Bjørn et al. 2018). Results can be displayed per impact category or summarised in a single environmental impact score, and the studies reviewed in this chapter have applied both approaches. While a single score will disguise differences and trade-offs between impact categories, it can serve as a valuable parameter to facilitate decision-making. The international recognition of LCA as an established methodology for environmental impact assessment is reflected in the norms ISO 14040 and 14044 (ISO 2006a, b). An LCA consists of four stages, which are goal and scope definition, life cycle inventory phase, life cycle impact assessment and interpretation of the results. The goal and scope of an LCA describe the processes considered within the system boundaries of the study and define the functional unit (FU) or reference flow as per which environmental impacts are calculated. For an LCA in the water sector, this could be $1\,m^3$ of water supplied, for instance. The following phase explains the data and methods used to calculate the environmental footprint. The third phase contains the results of the environmental footprint of the system for one or several environmental impact categories. In the last step, the results are interpreted to draw conclusions and give recommendations.

1.2.2 SELECTION METHOD FOR REVIEWED ARTICLES

In order to get a full overview of the literature on LCA in rainwater harvesting, a systematic search was conducted. This included the use of the scientific search engines Web of Science and ProQuest, which were screened using the search terms "rainwater" or "rain water" combined with the following terms: life cycle or lifecycle, greenhouse gas, CO_2, carbon emission, ecological footprint, ecological performance, environmental footprint, environmental performance and carbon footprint. After the removal of all duplicates, 1,096 articles remained, which were first reviewed by title and then by abstract. One hundred and two articles eventually remained for a full text review as they clearly contained LCA studies in the area of rain and/or stormwater systems. Studies that evaluated RWHS only on an inventory level, such as quantifying life cycle energy or water use, without determining the associated environmental impacts were excluded, as they do not fall under the LCA definition. Further, studies that did not separately report the environmental impacts of an RWHS but where the RWHS was only part of a larger infrastructure were excluded, too, as no conclusion to the RWHS itself could be drawn. While due to practicality and less pollutants, the typical catchment area for rainwater harvesting systems is roof tops (Yan et al. 2018), there are also systems that harvest runoff from roads and side-walks via (im-)permeable pavements (Antunes et al. 2020; Faragò et al. 2019). These could technically be regarded as a stormwater management strategy, but examples of this approach have been included here if the subsequent use of the harvested water was just like in a classic RWHS. For this review, 12 studies were reviewed in depth to showcase the breadth of the scope of the LCA studies on RWH, which can differ in the use of the rainwater, the alternative water system they are replacing, the layout and materials used or the number of households supplied. At the same time, all sampled studies in the review were deemed comprehensive and fulfilled minimum quality standards for conducting an LCA in regard to system boundaries, documentation of the data sources and inventory used, the use of well-known life cycle impact assessment methodologies, the use of standard software and/or databases, transparency, and disaggregation of results.

1.3 FROM ROOFTOPS TO PAVEMENTS: A REVIEW OF SELECTED LCAS ON RAINWATER HARVESTING

First, studies of different RWHS designs or materials are reviewed, exploring the environmental hotspots of RWHS. Next are the studies that compare RWH to other water supply options, such as the centralised municipal supply system or other alternative sources such as desalination or greywater recycling. Finally, a section covers literature on RWHS using paved areas and RWH for food supply in urban or rural agriculture. Table 1.1 provides an overview of the reviewed LCA studies and their key content.

TABLE 1.1

Overview of Scope and Results of the LCA Studies on RWH Reviewed in this Chapter

Reference	Country	Type of Area	Collection Area and Building Type (If Applicable)	Scale of RWHS	Use of Rainwater	Research Objective	No of Scenarios Assessed	System Boundary	Result	GWP Result	Unit
Angrill et al. (2012)	Spain	Residential	Rooftop; single-family houses and apartment blocks	1–10 buildings, diffusely or densely built	Laundry	Compare RWHS with varying tank locations and dwelling types	8	Cradle-to-grave (excl EoL of materials)	Best performance: gravity fed system with tank distributed over the roof. Gravity-fed RWHS in compact neighbourhood could compete with municipal water supply.	0.68–3.71	kg CO_2 eq/m^3
Angrill et al. (2017)	Spain	Residential	Rooftop; apartment block	One building, different number of floors considered	Laundry	Compare RWHs varying in tank location, building height and distribution strategy in building	24	Cradle-to-grave (excl EoL of materials)	Best performance: roof-top tank (gravity fed) system connected to communal laundry room as opposed to distribution to all individual apartments	0.4–0.5	kg CO_2 eq/m^3

(Continued)

TABLE 1.1 (*Continued*)

Overview of Scope and Results of the LCA Studies on RWH Reviewed in this Chapter

Reference	Country	Type of Area	Collection Area and Building Type (If Applicable)	Scale of RWHS	Use of Rainwater	Research Objective	No of Scenarios Assessed	System Boundary	Result	GWP Result	Unit
Antunes et al. (2020)	United Kingdom	Residential	Permeable pavement	City. 1 tank per approx. 600 inhabitants.	Toilets and outdoor use	Compare RWH with conventional municipal water supply and stormwater treatment	2	Cradle-to-grave (excl transport of materials)	RWH with permeable pavement better than conventional system in most categories and in the single-score result	21%	Lower GWP through RWH
Faragò et al. (2019)	Denmark	Residential	Rooftop and impervious surfaces; single family houses and housing units	City district	Toilet and laundry	Compare municipal groundwater supply to centralised and decentralised rain/ stormwater harvesting	5	Cradle-to-grave	Best performance: centralised rain/ stormwater harvesting with UF plus UV or H_2O_2 treatment. Rainwater option mostly better than conventional municipal supply	50%	lower GWP through best performing RWH (compared to municipal supply)

(*Continued*)

TABLE 1.1 (*Continued*)

Overview of Scope and Results of the LCA Studies on RWH Reviewed in this Chapter

Reference	Country	Type of Area	Collection Area and Building Type (If Applicable)	Scale of RWHS	Use of Rainwater	Research Objective	No of Scenarios Assessed	System Boundary	Result	GWP Result	Unit
Ghimire et al. (2014)	USA	Residential and agricultural	Residential: rooftop, single family house; agricultural: upland catchment, farmland	One building; 0.34 km² farmland	Residential: toilet; agricultural: irrigation	Compare pumped and gravity-fed domestic RWH to municipal supply for a single family house; and agricultural pumped or gravity-fed RWH to well water irrigation for a farm	6	Cradle-to-grave	Best performance: RWHS designs without pumping (Domestic and agricultural)	0.41 (domestic), 0.084 (agricultural)	kg CO_2 eq
Hofman-Caris et al. (2019)	The Netherlands	Residential	Single house: rooftop; city district: paved and built surfaces	One building and city district	Potable	Compare rainwater for potable use in densely populated city district or rural single house with central municipal supply from surface water	6	Use phase only	Best performance: whole district scale RWHS. RWH on both scales better than municipal supply	0.002 kg CO_2/m³ (operational only; Only CO_2)	kg CO_2 eq

(Continued)

TABLE 1.1 (*Continued*)
Overview of Scope and Results of the LCA Studies on RWH Reviewed in this Chapter

Reference	Country	Type of Area	Collection Area and Building Type (If Applicable)	Scale of RWHS	Use of Rainwater	Research Objective	No of Scenarios Assessed	System Boundary	Result	GWP Result	Unit
Leong et al. (2019)	Malaysia	Residential and commercial	Rooftop, domestic or commercial building	One building	Toilets and irrigation	Compare RWH with greywater harvesting (GWH), hybrid rain/grey water harvesting and municipal water supply	4	Cradle-to-grave (No on-site construction activities, no EoL)	Best performance: RWH (domestic), hybrid rain/greywater (commercial)	10% (domestic); 2% (commercial)	GWP savings compared to municipal supply
Prouty and Zhang (2016)	Uganda	Residential	Rooftop; single-family house	One household	Potable	Compare RWH with tap water, boreholes, springs and surface water	5	Cradle-to-grave	RWH (when appropriately treated) worst as needs most treatment	n.a.	
Sanjuan-Delmas et al. (2018)	Spain	Educational	Rooftop; single five-storey building	Supplying one rooftop greenhouse with 900 m² harvesting area	Irrigation	Compare rooftop greenhouse tomato cultivation (incl RWH) with conventional tomato cultivation	2	Cradle-to-grave	Better rooftop greenhouse than conventional agriculture	n.a.	

(Continued)

TABLE 1.1 (Continued)
Overview of Scope and Results of the LCA Studies on RWH Reviewed in this Chapter

Reference	Country	Type of Area	Collection Area and Building Type (If Applicable)	Scale of RWHS	Use of Rainwater	Research Objective	No of Scenarios Assessed	System Boundary	Result	GWP Result	Unit
Tarpani et al. (2021)	Brazil	Residential, educational, commercial, offices	Rooftop; single-household or apartment buildings, offices, schools, petrol stations	One building	Non-potable	Five types of RWH compared with desalination and indirect WW reuse after aquifer recharge	7	Cradle to grave	No clear result. RWH best in half of categories. RWH can compete with other alternatives if different tank material was used.	0.59	kg CO$_2$ eq
Valdez et al. (2016)	Mexico	Residential, commercial and office	Rooftop; single-family house, 3–7 storey residential buildings, 2–10 storey office building, 1–5 storey commercial building	One building	Potable and non-potable	Compare four types of RWHS varying in tank location (amongst others) with municipal supply	55	Cradle-to-grave (excl EoL of materials and construction activities)	All RWHS perform better. The higher the demand which could be met by rainwater, the lower the impacts for RWH	1.68–2.04	kg CO$_2$ eq
Yan et al. (2018)	United Kingdom	Educational	Rooftop; office building	One building	Potable	With mains water and bottled water	6	Cradle-to-grave		0.26–3.99	kg CO$_2$ eq

1.3.1 OPTIMISING DOMESTIC ROOFTOP RWHS THROUGH MATERIAL-EFFICIENT AND GRAVITY-FED CONFIGURATIONS

This section reviews and compares the environmental impacts of different types of rainwater harvesting systems. It focuses on typical rooftop collection systems and rainwater use for domestic purposes.

1.3.1.1 Tanks Distributed over the Roof and Compact Dwelling Infrastructure Enable Low-emission RWH for Laundry Use in Barcelona

The first study is the LCA of eight different scenarios for the design of rooftop RWHS in residential areas in Barcelona. It considered rainwater to be used for laundry and took into account the rainfall characteristics of the Mediterranean area (Angrill et al. 2012). The scenarios looked at various tank locations, such as underground, below-roof, rooftop and block tanks. Two different dwelling structures were considered which were compact urban areas featuring apartment houses and more diffusely built single-family houses with a garden. The seven impact categories – Abiotic Depletion, Acidification, Eutrophication, Global Warming, Human Toxicity, Ozone Depletion and Photochemical Ozone Creation – were analysed. The system boundaries comprised all resources and energy required from cradle-to-grave, including infrastructure and its demolition, but excluding the EoL of the materials themselves. For both the compact and the diffusely built dwellings, the LCA results identified that the RWHS with tanks distributed over the roof had the lowest impact in almost all of the impact categories. The compact dwelling structure, however, enabled an RWHS design with lower impacts per m^3 of rainwater harvested than the diffusely built neighbourhood, with the lowest determined GWP value of 0.68 kg CO_2 eq/m^3. The tank distributed over the roof had the lowest impacts because the design did not require a dedicated water catchment component, with the tank serving as a catchment area. Also, less material was required for the tank because the roof of the building functioned as the base of the tank and because an evenly distributed tank weight required fewer reinforcement structures for the building. Additionally, the operational savings were achieved as the system was gravity-fed. The main driver of the environmental footprint was the infrastructure of the RWHS. This was due to the extraction and manufacture of concrete and steel used for tank and reinforcement structures, especially the cement component of concrete and the transport of concrete components. The study also calculated the water volume required for the life-cycle of the infrastructure in a water footprint. Due to lowest material requirements for the tank-distributed over-roof in the compact urban neighbourhood, this scenario was the most water efficient, with a 1.5 m^3 water footprint per m^3 of water supplied, while other scenarios reached up to 17.7 m^3. Another difference between the two dwelling structures was the self-sufficiency for laundry use through rainwater. The larger catchment (roof) area available per inhabitant in the diffusely built neighbourhood meant that the rainwater met 98% of the laundry water demand there, as opposed to 47% in the apartment blocks. The use of the overflow of unused rainwater in the single-family houses, which was not considered in their study, could further minimise the impact of RWH per m^3 of water collected.

In a follow-up study, Angrill et al. (2017) investigated the use of rainwater specifically for newly constructed apartment buildings in a Mediterranean climate. This time, varying tank locations (underground or distributed over the roof), different building heights (number of apartments and inhabitants served from 30 to 84 apartments), and the distribution strategy for laundry use in the building (shared laundry room, supply to the nearest apartments only or distribution through the whole building to all apartments) were assessed. The environmental footprint showed all the rooftop tank options to perform better than the underground tank options in all impact categories analysed, due to the pumping energy required for underground tanks. For both tank options, the distribution strategy to a single laundry room performed best. Scenarios with a laundry room and roof tank had a 33%–42% lower carbon footprint than scenarios with a laundry room and underground tank, depending on the height of the building. When laundry water was distributed to all apartments individually in a 15-story building, roof tank options scored 61%–89% better, depending on the impact category. In most scenarios, the largest contributor to the carbon footprint was the material required to reinforce the building to withstand the tank's and the water's weight. However, compared to the construction material of the whole building, the reinforcement structures were negligible. The study also analysed the influence of climate regions found in Spain, other than the Mediterranean area, on the proposed RWHS. It found that the demand met with the same tank size would be significantly different and/or a different tank size would be required.

1.3.1.2   Maximising the Amount of Rainwater Used Brings Benefits despite Treatment to Drinking Water Quality

A study by Valdez et al. (2016) analysed the potential environmental benefits of four different types of RWHS for Mexico City, adapted to 11 building types. Those included residential, office and commercial buildings of different sizes with varying numbers of floors and occupants. RWH was deemed a promising solution for the city's water management issues. At the time of writing, Mexico City used water from an aquifer that had been overexploited for 50 years. This was supplemented by water imported from another basin, which was highly energy-intensive due to pumping. Additionally, an estimated 35% of water was lost through leakages before it reached the customer. The RWHS differed in the use type of the rainwater and tank location; two scenarios considered rainwater to be used for non-potable uses. One of the non-potable scenarios was gravity-fed with no pump and tanks located on the walls. Two other scenarios looked at potable rainwater use, where rainwater was treated via deep bed sand filtration and disinfected with silver ions. A special circumstance in the Mexican case study was that houses in Mexico City (even those without RWHS) were usually already equipped with certain water infrastructure: Water storage tanks, header tanks and pumps had been installed as the pressure in the municipal water system was not always sufficient. Therefore, when rainwater was used for potable use, the existing infrastructure could be shared between the municipal and the rainwater systems, lowering the additional amount of infrastructure required for an RWHS. The results for the RWHS included the operational savings from the avoidance of drinking water treatment and the avoided supply through the municipal system. The boundaries of the LCA included all infrastructure and energy demand from cradle to grave. Construction activities and disposal or recycling impacts from the material itself, however, were outside the scope of the study. Only global warming impacts were analysed. Generally, the RWHS, being able to supply a greater share of the total water demand, scored better. In the potable RWH scenarios, more rainwater could be used, as treatment of drinking water quality allowed it to be used for more purposes during seasons with plenty of rainfall, increasing rainwater use on an annual average. Rainwater could replace a maximum of 30% of the annual total water demand, depending on the building type and size. Another benefit was the reduction of required infrastructure, such as pumps and tanks, which could be shared with the municipal drinking water infrastructure. These benefits were higher than the additional impact from treating the rainwater to a potable standard. The RWHS with the lowest impact was one designed for a two-storey office building with a GWP of $1.68\,kg\ CO_2/m^3$ and sharing all infrastructure in the house with the municipal water.

1.3.1.3 Gravity-fed Outperforms Pumped Configuration Despite Increased Material Use

Ghimire et al. (2014) evaluated RWHS for both domestic use in a single-family house and agricultural use on a farm, located in Virginia, USA (see analysis of agricultural systems in Section 2.4.2). The domestic RWHS comprised a rooftop collection, a polyethylene (PE) underground storage tank, a submersible steel pump, pipes, gutters, filters and valves. The tank was designed to hold a volume large enough to supply a typical household in the Virginia Back Creek area with water for toilet flushing for two months. The pumped baseline system was compared to a minimal RWH solution with reduced pipework by placing a storage tank and toilet next to each other and a gravity-fed operation by elevating the tank with a steel structure. The LCA included cradle-to-grave life cycle impacts with infrastructure and energy requirements. In the pumped RWHS, pump infrastructure and pumping energy dominated the impacts, together with the PE tank, which was responsible especially for ecotoxicity and human health (cancer) impacts. The pipes made from PVC and CPVC particularly influenced ecotoxicity and smog formation. The minimal pump-less system scored better than the baseline pumped system in all of the 14 analysed impact categories, apart from metal depletion because of the additional steel requirements for the tank stand.

1.3.1.4 Centralised Rain and Stormwater System for the Danish Neighbourhood Offers Lower Environmental Impacts than a Decentralised System

Since climate change adaptation policies in Denmark require cities to plan for the mitigation of flood risks or combined sewer overflows, Faragò et al. (2019) assessed a rain and stormwater harvesting system for a new neighbourhood in Aarhus, Denmark. The neighbourhood is planned to accommodate 2,000 inhabitants and consists of about 80% single-family houses and 20% housing units. Five scenarios were analysed: one for conventionally centralised groundwater supply, three for centralised rainwater harvesting and one for decentralised rainwater harvesting. The rain or stormwater was intended to be for non-potable use, i.e., toilet flushing and laundry. The centralised RWHS collected stormwater from all available impervious areas (roofs and roads) and stored and distributed it through wet basins and channels. It applied different water treatment strategies in a central treatment plant: ultra-filtration (UF) with UV, UF with H_2O_2 or RO with UV. After treatment, water was stored in stainless steel tanks. The decentralised RWHS instead collected rainwater from the rooftop only and stored it in a PVC underground tank. The cradle-to-grave LCA encompassed all materials, including infrastructure. The most environmentally favourable RWH options were those designed in a centralised way, using UF combined with either UV or H_2O_2 as treatment. They scored best out of the RWH scenarios in three categories (global warming, fossil resource depletion and ecotoxicity) and equal in one (freshwater depletion) out of the six analysed impact categories. Their impacts were dominated by electricity consumption for water treatment and distribution (especially global warming), transport of material for basins and trenches (fossil resource depletion), the water treatment infrastructure (reserve base resource depletion[1]), pollutant runoff from storm/rainwater (ecotoxicity) and basin construction (particulate matter). Treatment with RO caused relatively higher environmental impacts, especially global warming, particulate matter and fossil resource depletion, due to the high electricity requirements at the treatment stage. In the decentralised RWHS, global warming impacts were caused by 72% of rainwater pumping from the underground storage tank. Ecotoxicity was higher in the centralised systems, as roof runoff was considered to be discharged to the WWTP either directly as overflow or after use in the household. In the centralised RWHS, instead, some pollutant removal has already occurred in the basins and trenches.

1.3.1.5 Tank Manufacture Dominates Environmental Impacts for Non-potable Rainwater Supply in Brazil

Tarpani et al. (2021) explored several water options for the northern part of the city of Florianopolis in Brazil to enhance or compliment the current municipal drinking water supply system. It investigated RWH amongst other water strategies (see Section 2.3.2). The RWHS were designed to fit five types of buildings: single households, residential buildings, offices, schools, and petrol stations. They supplied water for non-potable uses such as toilet flushing, laundry or car washing. Local rainfall, the available catchment area and suitable storage tank sizes were considered. The system contained a storage tank on ground level, from which water was pumped to an elevated header tank where water was disinfected via UV. The cradle-to-grave LCA included all infrastructure, operation, and disposal and analysed the impacts in 12 categories. The carbon footprint for RWH was calculated at $0.59\,kg\,CO_2$ eq/m^3. Overall, the most important contributors to the impacts of RWH were the production of the glass fibre storage tanks, followed by steel for the pump and electricity for the UV lamps. For global warming, 75% was attributed to glass fibre for the tanks, mainly from energy-intensive nylon content production and direct emissions from nylon production. Glass fibre production, especially related energy consumption, also dominated impacts of RWH on water depletion and marine eutrophication. The authors therefore suggested the use of PE storage tanks as a measure to reduce the environmental impact. Differences between RWHS adjusted to different buildings were not discussed. The study, however, included sensitivity analyses by applying the electricity grid mixes of other countries, namely Mexico, Argentina and South Africa. RWH global warming impacts ranged from 591 kg CO_2 eq/1,000 m³ for Brazil, which used 63% hydropower in its mix, to 1,878 kg CO_2 eq for South Africa, using 92% coal. These variations nicely show the influence of geographic differences and the difficulty of transferring footprints and findings between countries.

1.3.2   RWH vs. Other Supply Options – Benefits and Trade-offs

1.3.2.1   RWH Compared to Conventional Municipal Supply

1.3.2.1.1   Benefits from RWH Depend on Infrastructure, Energy Requirements and Water Hardness

Valdez et al. (2016) regarded RWH as an addition to the municipal groundwater supply of Mexico City and considered avoided burdens from the municipal system as credits to RWH. They found all RWHS studied to lower the GHG emissions of the supply and treatment of 1 m³ of water. This was independent of the RWHS configuration, such as pumped or gravity-fed, and despite the fact that infrastructure requirements were higher for RWH. In high-rise buildings, GWP savings were highest for the gravity-fed system, with 3%–9%. However, in low-rise buildings, the largest savings of up to 18% could be achieved through a system for potable use, sharing essential equipment within the building for municipal supply. In Ghimire et al.'s (2014) study on Virginia, the comparison between municipal and rainwater supply depended on the RWHS configuration. Only when the RWHS was gravity-fed and used reduced pipework, did the domestic RWHS system reduce environmental impacts compared to the municipal supply in all 14 impact categories except the eco-toxicity impact category. This was despite the necessity for additional infrastructure to elevate the tank for gravity operation. The reduction in GWP was 52%, and energy consumption was 58% lower.

The comparison of the LCA results for centralised and decentralised RWH with groundwater municipal supply in Aarhus was also case-dependent (Faragò et al. 2019). Most rainwater solutions offered overall environmental benefits compared to the municipally centralised water supply strategy. The centralised rainwater supply options with UF treatment proved to reduce environmental impacts in all impact categories except reserve base resource depletion. Regarding ecotoxicity, the centralised rainwater options scored better because the centralised treatment of the stormwater removed pollutants, which would not be the case in the municipal system. The centralised rainwater option with RO treatment scored better than conventional supply only in terms of ecotoxicity and freshwater depletion. The decentralised rooftop rainwater scenario offered benefits in the particulate matter, reserve base resource depletion and freshwater depletion categories, i.e., half of the assessed categories. Interestingly, Faragò et al. (2019) also considered the effect of the different hardness levels of rainwater and groundwater. The higher hardness level of the municipal water meant that electricity consumption and laundry detergent use were higher in this scenario, which resulted in 22% higher GWP values for municipal groundwater compared to rainwater. Similarly, a study by Vargas-Parra et al. (2019) on rainwater use for laundry purposes in Barcelona/Spain, found huge benefits from softer rainwater. Here, savings from avoided detergent use surpassed the environmental impacts from the RWHS more than 20-fold in all but one of the six analysed impact categories.

Angrill et al. (2012) compared the carbon footprint per m^3 of RWH in their study with literature values for conventional municipal supply in the Mediterranean climate. They concluded that only the gravity-fed RWHS of their study, with a footprint of 0.64 kg CO_2 eq/m^3, could compete with municipal water supply, reaching similar GWP values.

1.3.2.1.2 Rainwater for Potable Use on a University Campus in the UK Cannot Cope with Municipal Supply Due to Energy-intense Treatment Method

Yan et al. (2018) compared RWH exclusively for potable use with municipal water supply to a university office and conference building in the United Kingdom (UK). Rainwater was harvested from the roof, treated to a potable standard, stored in an underground tank, pumped and delivered to users in the building. A new point-of-use (POU) treatment device was tested for the system, comprising a 5 μm inlet filter, UV, ozone, and a carbon filter. The system boundaries included all steps from manufacture to operation, maintenance, and disposal of the POU device, including the electricity consumption to run the treatment device. The comparative system, the centralised conventional water supply system, was considered with the manufacture of the infrastructure, water treatment and distribution steps. The LCA results revealed that the rainwater supplied via the POU device had a higher environmental footprint in all 18 accessed impact categories than the conventionally centralised supply. In terms of the carbon footprint, rainwater caused 3.99 kg CO_2 eq/m^3 supplied versus 0.44 kg CO_2 eq/ m^3 supplied by the municipal system. This was due to the large electricity consumption, both directly from the operation of the POU device as well as indirectly from the manufacture of the device, which was run with UK grid electricity. This contributed especially to the high impacts in the freshwater ecotoxicity, agricultural land occupation and water depletion categories. The authors included scenarios with other electricity sources, such as photovoltaic (PV), onshore wind and the UK's anticipated grid mix for 2030. Changing the RWH electricity source to renewables could partially provide a benefit for RWH. In the case of wind energy, RWH scored better in 11 impact categories, namely climate change, terrestrial acidification, photochemical oxidant formation, particulate matter formation, terrestrial ecotoxicity, ionising radiation, agricultural land occupation, urban land occupation, natural land transformation, water depletion and fossil resource depletion. In the case of wind energy, RWH showed environmental benefits only for urban land occupation and natural land transformation. Nevertheless, in 2030, when both mains water supply and the RWH option would use a mix with more renewables, the benefit of using mains water would remain.

1.3.2.1.3 Dutch District RWHS for Potable Use Scores Better than Municipal Supply

Hofman-Caris et al. (2019) presented another study looking into the use of rainwater as a drinking water source, taking Amsterdam as an example. The scenarios in their study explored the environmental feasibility of RWH in either a newly built, densely populated city district where rainwater is collected via paved and built surface areas or at a single peri-urban house from its rooftop. The six scenarios further varied in terms of collection surface area, storage system – open concrete pond or closed HDPE tank – and treatment technology. All scenarios involved the pumping of the water. The two main treatment options considered were either via RO filtration combined with UV disinfection or via sand filtration followed by UV/H_2O_2 and activated carbon. In contrast to the previous studies described in this chapter, only the operational expenses (use phase) were included in the LCA, such as energy for pumping and treating, as well as chemicals and materials (e.g., H_2O_2, activated carbon, sand, etc.). RWH was contrasted with the conventional centralised drinking water supply from surface water, again in the operational phase with no infrastructure. The environmental impact results were shown as either "ecopoints" (measured in milli ecopoints, mPT), which is a single score that combines several impact categories, or as CO_2 footprint (CO_2 only – no other GHGs). The two RWH district scenarios scored considerably better (15 and 12 mPT/m^3) than the central drinking water supply (36 mPT/m^3). The single-house RWH scenarios also showed values lower than central drinking water, with UV/H_2O_2-scenarios being better (24 mPT/m^3) than RO-scenarios (33 mPT/m^3). In terms of operational CO_2 emissions, the RWH scenarios also scored considerably better than the central drinking water supply. The reason for the higher environmental impact of the municipal supply were found to be long transport distances of 60 km from the abstraction point, treatment through coagulation, and the necessity for softening. All these process steps were not necessary for RWH. However, the authors did not recommend RWH for the following reasons: apart from rural locations, there was no self-sufficiency from rainwater and the municipal central supply was still required; infrastructure, which has been shown in other studies to increase the impact per m^3 greatly, was not considered in the study; and the environmental benefit of RWH was negligible (about 1%) if compared to the total environmental impact of a person per year. Furthermore, the environmental impact of RO concentrate disposal and treatment was not considered.

1.3.2.2 RWHS Compared to Other Alternative Water Supplies

1.3.2.2.1 Clear Benefits from RWH vs. Desalination and Water Reuse for Carbon Emissions Only

The study on alternative supply options to the current municipal system in Florianopolis compared other possibilities in addition to RWH (Tarpani et al. 2021). One was the desalination of seawater via RO for drinking water delivered to the existing water supply network. The other was the indirect potable reuse of wastewater, where wastewater was first treated via anaerobic digestion, oxidation and ozonation, then infiltrated into the aquifer (aquifer recharge), pumped into the existing network and finally chlorinated. RWH outperformed seawater desalination and indirect wastewater reuse in 6 out of the 12 impact categories, which were global warming, ozone depletion, freshwater eutrophication, ionising radiation, terrestrial ecotoxicity and human toxicity. However, it caused the highest impacts of all three solutions in water depletion and marine eutrophication and had intermediate results in the remaining categories. In global warming, RWH performed best with 0.59 kg CO_2 eq./m^3 compared to 0.64–0.87 kg CO_2 eq. for seawater desalination and

$0.87-1.12$ kg CO_2 for indirect wastewater reuse depending on electricity consumption. However, as RWH was the least electricity-intensive system in their comparison, the country's electricity grid mix had a smaller influence on the environmental impact than in the case of desalination and wastewater reuse. Therefore, RWH impacts were more transferrable to other countries. A similar study (Godskesen et al. 2011) compared RWH for toilet flushing with brackish water desalination for potable use and artificial aquifer recharge with lake water for Copenhagen/ Denmark. It found RWH to have the least environmental impact but was unable to provide water self-sufficiency.

1.3.2.2.2 Hybrid Rainwater-Greywater System Maximises Savings on Municipal Water

Motivated by recent water shortages in parts of Malaysia, Leong et al. (2019) studied a hybrid rainwater-greywater system (HRG) that uses rainwater during wet monsoons and greywater during dry monsoons, thus maximising water savings of potable municipal water. The hybrid system was compared to the centralised municipal water supply from surface water and to two decentralised systems: decentralised RWH and decentralised greywater harvesting (GWH; treatment with sand filter, granular activated carbon (GAC) filter and ozone disinfection). All four scenarios were considered to supply non-potable water for toilets and irrigation, then evaluated for both a five-person household and a commercial site. The system boundaries included all life cycle stages apart from EoL, including the use phase and related energy requirements. Infrastructure such as pipework and tanks were included, apart from construction on-site. All three alternative water supply options (RWH, GWH and HRG) could switch to municipal supply if no water was available through the other route. The RWHS comprised a rooftop collection area and tanks from PE, but no water treatment system. In the HRG option, greywater was treated via sand and GAC filters and disinfected with ozone. The hybrid scenario contained separate equipment for both RWH and GWH, such as separate tanks.

The domestic and commercial settings differed in the environmentally most favourable option. In the commercial building, the hybrid system scored best in six out of the nine impact categories analysed, including global warming. Only in abiotic depletion, ozone layer depletion and water stress did other options score better. And only for abiotic depletion it scored worse than the municipal supply. The stainless-steel stands for the tanks were responsible for high impacts in the abiotic depletion and ozone layer depletion categories. The hybrid system saved most municipal potable water by being able to provide 55% of the water demand for toilet flushing and irrigation (vs. 19% with RWH only). The main environmental benefits of the hybrid system were caused by avoided impacts from the municipal water system, both avoided water treatment and supply as well as avoided wastewater treatment. Avoided wastewater specifically reduced impacts in the eutrophication and ozone depletion categories.

For the domestic building, RWH turned out to be the most advantageous option. It scored best compared to all other options in four impact categories and better than the municipal system in seven out of nine categories. Rainwater could substitute for 95% of the non-potable water demand. Due to the lower water demand in the domestic setting, most of the water in the hybrid system was rainwater. The grey-water recycling system instead was under-utilised and could avoid only little waste-water and related impacts. In comparison, the hybrid system could supply 100% of non-potable water use, not a big difference from the RWHS.

A sensitivity analysis varying the energy intensity (kWh/m³) for all four systems by ±20% showed only minor effects on the results. The global warming impact in the GWH system for the commercial setting changed by only ±3%.

1.3.2.3   Influence of User Behaviour on the Environmental Profile of Rainwater Harvesting

Prouty and Zhang (2016) brought a new dimension to LCAs on water supply with a study on the environmental impacts of drinking water use in Uganda, considering people's perceptions of the quality of different water sources. The inhabitants of two rural villages in the study area in central Uganda had access to different water sources, which were compared in the study: tap water, boreholes, springs, rainwater harvested from the roof and surface water primarily from Lake Victoria. The system boundaries of the LCA of the water supply options comprised the water source, water collection, household treatment and use of the water, including (where applicable) infrastructure, transport and required energy. For example, rainwater was harvested from roofs and stored in recycled metal oil drums or locally fabricated metal or plastic tanks. Two scenarios were compared for each water supply option: the life cycle environmental impacts when the water was treated by the household according to the measured water quality; and the actual impacts the water supply options currently have due to the treatment based on perceived water quality. Results were shown for twelve impact categories.

Household surveys revealed that harvested rainwater was not treated by households at all despite having the worst water quality with high levels of pathogens such as *T. coliform* and *E. coli*, turbidity and total dissolved solids (TDS) due to the lack of a first flush. This resulted in rainwater having the lowest actual environmental impact of all water sources. However, based on the measured contaminants, the recommended treatment would have included settling, filtration, boiling and chlorination.

Tap water, in contrast, although fulfilling WHO drinking water standards and not requiring treatment by households, was boiled by 90% of households using charcoal, making it the environmentally worst option due to perceived low water quality. The combustion of charcoal has especially contributed to impacts in the land use and global warming categories. If rainwater had been treated to drinking water standards, it would have had the highest environmental impacts of all options, closely followed by surface water requiring similar treatment. The most relevant impact categories contributing to the end-point results for both rain and surface water when treated as recommended would be land use and global warming due to the burning of charcoal.

The study showed how perceptions of water qualities and associated user behaviour can lead to inadequate water treatment practices and unexpected environmental performance of water supply options. As inadequate treatment also causes potential health risks, the authors called for educational efforts to foster public awareness of the issue.

1.3.3 RWH BEYOND THE CLASSICAL ROOFTOP SYSTEM

1.3.3.1 RWH from Pavements

1.3.3.1.1 Centralised Rainwater Harvesting Systems Have Lower Environmental Impacts than Municipal Groundwater Supply

The study for Aarhus included a typical rooftop RWHS as well as harvesting from paved areas such as streets or sidewalks (Faragò et al. 2019). To help prevent floods, the three centralised RWHS for a whole district collected and distributed rainwater through open trenches and channels and stored it in surface basins; 76% was collected from roofs and 24% from streets and other impervious surfaces. Despite higher infrastructural requirements for the central treatment plant, such as steel for storage tanks, piping and cleaning equipment, the two centralised RWHS (which did not use RO for water treatment) had overall lower environmental impacts than municipal groundwater supply. Only in reserve-based resource depletion did they score worse. Steel tanks contributed 24%–29% to that category in all centralised RWH scenarios, and basin and trench material contributed over 40%.

1.3.3.1.2 Permeable Pavement System Deemed Beneficial for Newly Built Residential Area in Glasgow

Antunes et al. (2020) looked at rainwater collection through permeable pavements as part of a sustainable drainage system, a collection method typically used for stormwater management. The study, which took the city of Glasgow/UK as a case study area, quantified the environmental implications of collecting rainwater from roads and sidewalks. Subsequently, the water was distributed to residential buildings and chlorinated for non-potable use in toilets and for outdoor uses. Filtration through the permeable pavement was considered sufficient to hold back sediments. While the authors could not rule out the necessity of further filtration or treatment in practice, it was not considered in the study. One underground tank of 500 m³ capacity was assumed to serve about 600 people. The sizing was determined via the collectable rainwater amount based on rainfall and paved surface area as well as the reported demand. The permeable pavement system was compared to a conventional stormwater management system and a municipally centralised water supply system. That is, for both the permeable pavement and the conventional system, cradle-to-grave life cycle impacts were determined, from raw material extraction and building of the infrastructure through maintenance, operation (incl. electricity for pumping) and final disposal. Both systems were compared as newly built as opposed to retrofits. For the traditional system, this included conventional streets with impermeable asphalt.

For 92% of the days, the water demand for toilet flushing and outdoor uses could be completely met. The permeable pavement option showed an improvement in terms of environmental impact in all but four of the 15 analysed impact categories, e.g., a 20% reduction in GWP. The highest environmental savings were shown through the permeable pavement system in the water consumption category (99%), as this also accounted for the savings of potable water replaced through rainwater. The only significant environmental disadvantage was observed in the fossil resource scarcity category, with 20% higher impacts for rainwater collection due to higher material requirements for pipes and tanks, etc., and the use of petroleum for their manufacture.

Grouping the impacts into the three so-called endpoint damage categories of human health, ecosystem and resources showed benefits of the rainwater scenario in human health and ecosystem impacts and a higher impact on resource consumption. However, weighting and summarising to a single score revealed a 47% overall improvement in environmental impact through RWH. The weighted impact categories with the highest impact and most important reductions through RWH were water consumption, particulate matter formation and global warming. The conventional pavement and drainage system had higher impacts on water and particulate matter because of the heat required to produce the hot mix asphalt used as the pavement surface.

The authors concluded that a permeable pavement system with the collection and use of stormwater was advantageous for a city like Glasgow if built new. A change from the conventional to the proposed system was deemed unlikely, and demolition of the old system was not considered in the comparison.

1.3.3.2   RWH for Irrigation in Field-based and Urban Agriculture

*1.3.3.2.1   Gravity-fed Rainwater Irrigation System Yields Savings
in All Assessed Environmental Impacts*

While so far this chapter has discussed studies of domestic use of rainwater, LCA has also been applied to RWH in agriculture. In Ghimire et al. (2014), rainwater supply was compared to well water irrigation for a 0.34 km^2 farm in Virginia/USA. The agricultural system considered the irrigation requirements of a corn crop for a year with typical rainfall. It comprised a catchment area, channel, sedimentation chamber, PE tank, irrigation equipment, pipes, valves, and filter. Similar to the domestic system, they analysed both a pumped and a gravity-fed RWHS. The pumped agricultural RWHS scored better than the well water system in all but three impact categories, and the minimal design without pump in all impact categories. In the latter case, GWP savings of 76% were reached. Large reductions were also found for ozone depletion (67%), acidification (87%), smog (80%), ecotoxicity (72%), eutrophication (77%) and human health impacts (58%–67%), depending on the human health category. This was predominantly achieved through energy savings of 78% compared to the well water system. In the categories of metal depletion and non-cancer related human health impacts, large savings were additionally reached through the avoided pump manufacture.

*1.3.3.2.2   Rooftop Greenhouse with Rainwater Irrigation More Sustainable
than Field Cultivation Due to Local Supply Scheme*

Sanjuan-Delmás et al. (2018) analysed the environmental impacts of a system that brings agriculture into the urban space, looking at a rooftop greenhouse (RTG). It used rainwater for irrigation in a Mediterranean climate, as exemplified in Barcelona/Spain.

The greenhouse was integrated with the building, i.e., synergies with the building existed in terms of resource use such as rainwater, residual energy (heat), or residual air (CO_2). For the case study, tomatoes were considered the crop grown in the greenhouse, and LCA results were compared to conventional tomato cultivation. The LCA considered the complete system from cradle to grave, including operation and EoL, i.e., demolition of the whole system. Primary data could be used from the integrated greenhouse installed on the top floor of a Barcelona university building. The RWHS collected water from the roof, which was filtered to remove solids and then stored in a 100-m³ underground tank, from where it was pumped to the plants on the fourth floor. The water was used for the tomatoes as well as the ornamental plants in the building.

Results showed that overall, rooftop tomato cultivation had lower environmental impacts in five of the six analysed impact categories, e.g., 66% lower GWP impact per kg of tomato than tomatoes from conventional agriculture. This was to a large extent caused by the avoidance of packaging and distribution of the produce in the greenhouse case, where tomatoes are consumed directly and locally by the users of the building. RWH could satisfy 80%–90% of the irrigation water demand, depending on the season. RWH infrastructure contributed between 10% and 25% in four out of six categories, related to the construction of a 100 m³ tank from glass fibre-reinforced polyester. However, they found that by applying rainwater tank size optimisation software, 90% of the rainwater used could have been provided with a tank capacity of only 20 m³. A smaller tank size was therefore one of the optimisation suggestions for the rooftop greenhouse made by the authors. Impacts from the operation (energy use) for the RWH were not shown separately from other operational expenditures. Inventory data showed that the rainwater pump consumed between 0.8 and 1.5 kWh/ kg of tomato grown depending on the season, which was considerably less than the 9–17 kWh/kg consumed by the pump for the nutrient solution. It could therefore be expected that environmental impacts from rainwater pumping only contributed to the overall greenhouse operation to a minor extent. A rooftop tank location was not considered in the study, although rainwater was collected and mainly used on the rooftop.

1.4 DISCUSSION

The reviewed articles showed a variety of applications for RWHS: potable vs. non-potable use, domestic vs. agricultural use and a range of configurations: pumped vs. gravity-fed, different tank locations and materials, as an addition or a substitute for a conventional municipal system. They differed further in the application of the LCA methodology. While most included infrastructure from cradle to grave, some excluded life cycle stages such as construction activities or demolition (Leong et al. 2019) or excluded infrastructure altogether (Hofman-Caris et al. 2019). They also applied different impact assessment methods and impact categories, and GWP was the only category included in all. Environmental impacts were not always shown disaggregated per impact category but sometimes aggregated to a single environmental impact score, which is a valid method in LCA.

Factors of LCA methodology that influenced the results in the reviewed papers were found to be:

System boundaries: inclusion or exclusion of certain life cycle stages, such as infrastructure, transports or EoL.

Life cycle impact assessment: selection of impact categories, display of results per category or summarisation to a single score. Are potentially important impact categories left out? Are trade-offs between different categories revealed?

Completeness: The expansion of the LCA to include avoided impacts from avoided municipal or other alternative water supplies as credits for RWHS.

1.4.1 CONSIDERATE DESIGN CRUCIAL TO INCREASE ENVIRONMENTAL SUSTAINABILITY OF RWHS

Significant environmental impacts from RWH were both attributable to infrastructure such as tanks, holding structures for tanks or reinforcement structures for buildings where a collection or storage system was located on the roof or wall of a building. Major impacts were also caused by operational expenditures such as pumping rainwater when the system was not gravity-fed, as well as treatment energy or chemicals depending on the technology used and desired water quality.

It was case-dependent whether the manufacture of infrastructure or electricity consumption during the use of phase dominated the impacts. As an illustration, when comparing the GWP values for a gravity-fed system for domestic buildings in a Mediterranean city, impacts from the manufacture of infrastructure were reported as 87%–94%, followed by the transport of the material, while impacts from the use phase were negligible (Angrill et al. 2012). Despite the need for pumping, infrastructure dominated GWP impacts in the Florianopolis case with about 85%, mainly from glass fibre for the tanks, while only about 15% was attributable to electricity consumption for pumping and UV treatment during the operation (Tarpani et al. 2021). Also in the pumped domestic system for Virginia, infrastructure dominated, with 60% in GWP (Ghimire et al. 2014). In Angrill et al.'s study (2017), the materials needed to reinforce the building to withstand the tank's and water's weight contributed the most to GWP values in most of their scenarios for apartment buildings.

Other pumped systems showed a dominance of GWP impacts from use phase electricity consumption. The highest share of GWP impact from the use phase was observed for the potable system of the UK university office building with 96% (Yan et al. 2018), followed by 90% in the pumped agricultural system (Ghimire et al. 2014), the system with an underground tank for an apartment building with 78% (Angrill et al. 2012), and the two best performing centralised RWHS for Aarhus with 50% (Faragò et al. 2019).

The following strategies to improve RWHS environmental performance were suggested by the reviewed papers:

Consideration of tank location to enable gravity-fed operation as opposed to pumped systems (Angrill et al. 2012; Ghimire et al. 2014; Angrill et al. 2017)

Compact design of the distribution system to minimise impacts from pipework and pumping (Angrill et al. 2017; Ghimire et al. 2014)

Reduction of impact from tank materials by choosing the right (minimal) tank size (Sanjuan-Delmas et al. 2018)

Reduction of material impacts by using alternative, lower-impact materials for the tank (Tarpani et al. 2021) or integrating the collection and storage system with the building (Angrill et al. 2012).

Minimisation of operational electricity consumption through use of efficient pumps (Yan et al. 2018), considerate choice of the treatment technology (Faragò et al. 2019) and use of renewable energy (Yan et al. 2018).

While it may be feasible to implement the abovementioned measures for most RWHS, other factors that also influence the environmental footprint are more site-specific. There was a considerable influence of the dwelling type or dwelling density on how environmentally "efficient" an RWHS could be designed. Angrill et al. (2012) found benefits if systems were designed for higher population densities, as prevalent in apartment buildings. And the height of a building was also seen to influence environmental impacts, e.g., because of pumping energy (Angrill et al. 2017; Valdez et al. 2016). The balance between water availability and demand, determined by rainfall patterns, rooftop area, amount of users and type of water use (e.g. for non-potable uses only or all uses), influenced the environmental footprint as it determined the size of the tank in order to meet maximum self-sufficiency through rainwater (Angrill et al. 2017; Sanjuan-Delmas et al. 2018; Leong et al. 2019). The same factors determined the productivity of the RWSH, or "tank productivity", as Ghimire et al. (2014) called it, which is the water produced or rainwater used per volume of tank. In general, the more water that can be harvested AND used, the lower the environmental footprint per m³. At the same time, the more municipal potable water could be replaced by rainwater, the higher the potential for avoided impacts (Valdez et al. 2016). Electricity consumption impacts were highly influenced by the energy sources in the country and the share of fossil or renewable fuels (Tarpani et al. 2021). Cross-country comparisons for these impacts must therefore be taken with care, and for an ultimate evaluation of an RWHS, energy use in kWh should be added for a fairer comparison. Tarpani et al. (2021) compared the Florianopolis RWH electricity consumption of 0.38 kWh/m³ to values from different mains water supplies from Europe, the US and Canada, ranging from 0.4 to 1.71 kWh/m³ (Wakeel et al. 2016). They concluded that RWH can compete in terms of electricity consumption.

1.4.2 IS RAINWATER MORE SUSTAINABLE THAN OTHER WATER SOURCES? IT DEPENDS

The favourable comparison between RWH and other water supply options was also case-dependent. This was influenced both by the RWH system itself and also by the performance of other options such as municipal supply from ground or surface water, or supply from desalination, greywater harvesting or indirect potable reuse of wastewater. RWHS in this review showed a GWP impact per m³ of 0.26–3.99 CO_2 eq/m³ (Yan et al. 2018) (Table 1.1). GWP savings per m³ of water delivered using

rainwater ranged from negligible to 76% for the best performing RWHS. The highest
GWP reductions were found by replacing well water with rainwater in a gravity-fed
system in agriculture (76%) (Ghimire et al. 2014). About 50% reduction was achieved
through a decentralised domestic system without a pump (Ghimire et al. 2014) and a
centralised pumped system compared to conventional municipal supply. Other studies showed 20% GWP benefits through RWH from permeable pavements (Antunes
et al. 2020), 18% for a pumped system in low-rise buildings (Valdez et al. 2016), 10%
for a domestic system (Leong et al. 2019), or 9% for a gravity-fed system in high-rise
buildings (Valdez et al. 2016). RWH was also found to score better than desalination by 7%–32% and better than indirect reuse of wastewater by 32–47% in GWP
(Tarpani et al. 2021), but overall scored better only in half of the analysed impact categories. In other studies, RWH could also cause worse GWP impacts, as shown for
a centralised RWHS using RO for treatment, where GHG emissions were about 50%
higher than for the municipal system (Faragò et al. 2019). In the case of the RWHS
for potable use with a POU device, GWP impacts increased even by 800% (Yan et al.
2018). As the throughput through a water supply system determines the footprint per
m^3, extended centralised municipal systems generally have an advantage compared
to smaller or decentralised RWHS (Yan et al. 2018). For a fair comparison in this
regard, large-scale RWHS would have to be considered (Angrill et al. 2012). Finally,
user behaviour can change environmental burdens in an unforeseen way, as shown
for the Uganda villages (Prouty and Zhang 2016).

Even if out of the scope of this review, it should be mentioned that many of the
discussed studies also included economic assessments of the systems (Faragò et al.
2019; Valdez et al. 2016; Leong et al. 2019), which is a decisive factor for RWH
uptake, at least where other water supply options exist.

1.5 CONCLUSION

Overall, the environmental advantages and trade-offs of rainwater as an alternative
to municipal ground or surface water supply or other alternative water supply options
are case-specific, and there is clearly no one-fits-all solution. To analyse the full picture and not leave out significant impacts, the full life cycle of the system, including
infrastructure, should be considered, and the quantification of impacts should not
be restricted to GHG emissions. The design of the system is crucial to maximise its
environmental sustainability and minimise trade-offs between environmental categories and needs to consider building structure, population density, water demand
and annual and seasonal rainfall availability. A range of measures are available to
reduce the environmental impact of rainwater harvesting, including material choice,
gravity-fed operation, and compact and minimal sizing of tanks and pipes. A care-
fully designed RWHS takes into account low-impact materials and low operational
energy use and can provide a worthwhile strategy to diversify water supply in both
urban as well as rural and agricultural spaces, as it also comes with benefits usually
not covered by LCA's such as improved water supply security and flood prevention
(Valdez et al. 2016; Faragò et al. 2019). Rainwater harvesting can be a valid measure
to combat climate change impacts on watersheds as it reduces the stress on blue

water resources and can save energy for the treatment of drinking water standards or stormwater treatment. Whether it provides a benefit over other water supply options is case-specific, and all scenarios need to be thoroughly investigated before a final decision is made.

NOTE

1 Reserve base resource depletion defines the depletion potential of a resource, taking into account the relation between the extracted mass of a resource and its quantity in natural deposits (Klinglmair, Sala, and Brandão 2014).

REFERENCES

Amores, Maria José, Montse Meneses, Jorgelina Pasqualino, Assumpció Antón, and Francesc Castells. 2013. "Environmental Assessment of Urban Water Cycle on Mediterranean Conditions by LCA Approach." *Journal of Cleaner Production* 43 (March): 84–92. https://doi.org/10.1016/J.JCLEPRO.2012.12.033.

Angrill, Sara, Ramon Farreny, Carles M Gasol, Xavier Gabarrell, Bernat Vinolas, Alejandro Josa, and Joan Rieradevall. 2012. "Environmental Analysis of Rainwater Harvesting Infrastructures in Diffuse and Compact Urban Models of Mediterranean Climate." *International Journal of Life Cycle Assessment* 17 (1): 25–42. https://doi.org/10.1007/s11367-011-0330-6.

Angrill, Sara, Luis Segura-Castillo, Anna Petit-Boix, Joan Rieradevall, Xavier Gabarrell, and Alejandro Josa. 2017. "Environmental Performance of Rainwater Harvesting Strategies in Mediterranean Buildings." *International Journal of Life Cycle Assessment* 22 (3): 398–409. https://doi.org/10.1007/s11367-016-1174-x.

Antunes, Lucas Niehuns, Calum Sydney, Enedir Ghisi, Vernon R Phoenix, Liseane Padilha Thives, Christopher White, and Emmanuelle Stefânia Holdefer Garcia. 2020. "Reduction of Environmental Impacts Due to Using Permeable Pavements to Harvest Stormwater." *Water* 12 (10): 2840. https://doi.org/doi.org/10.3390/w12102840.

Bjørn, Anders, Mikolaj Owsianiak, Christine Molin, and Alexis Laurent. 2018. "Main Characteristics of LCA." In *Life Cycle Assessment - Theory and Practice*. edited by Michael Z. Hauschild, Ralph K. Rosenbaum, and Stig Irving Olsen. Cham, Switzerland: Springer International Publishing, 1st ed., 9–16. doi:10.1007/978-3-319-56475-3.

Campisano, Alberto, David Butler, Sarah Ward, Matthew J Burns, Eran Friedler, Kathy DeBusk, Lloyd N Fisher-Jeffes, et al. 2017. "Urban Rainwater Harvesting Systems: Research, Implementation and Future Perspectives." Water Research 115 (May): 195–209. https://doi.org/10.1016/j.watres.2017.02.056.

de Sá Silva, Ana Carolina Rodrigues, Alex Mendonça Bimbato, José Antônio Perrella Balestieri, and Mateus Ricardo Nogueira Vilanova. 2022. "Exploring Environmental, Economic and Social Aspects of Rainwater Harvesting Systems: A Review." *Sustainable Cities and Society* 76 (March 2021). https://doi.org/10.1016/j.scs.2021.103475.

Faragò, Maria, Sarah Brudler, Berit Godskesen, and Martin Rygaard. 2019. "An Eco-Efficiency Evaluation of Community-Scale Rainwater and Stormwater Harvesting in Aarhus, Denmark." *Journal of Cleaner Production* 219 (May): 601–612. https://doi.org/10.1016/j.jclepro.2019.01.265.

Friedrich, E., Shanthi Pillay, and C. A. Buckley. 2009. "Carbon Footprint Analysis for Increasing Water Supply and Sanitation in South Africa: A Case Study." *Journal of Cleaner Production* 17 (1): 1–12. https://doi.org/10.1016/J.JCLEPRO.2008.03.004.

Ghimire, Santosh R., and John M. Johnston. 2015. "Traditional Knowledge of Rainwater Harvesting Compared to Five Modern Case Studies." World Environmental and Water Resources Congress 2015: Floods, Droughts, and Ecosystems - Proceedings of the 2015 World Environmental and Water Resources Congress: 182–193. https://doi.org/10.1061/9780784479162.017.

Ghimire, Santosh R., John M. Johnston, Wesley W. Ingwersen, and Troy R. Hawkins. 2014. "Life Cycle Assessment of Domestic and Agricultural Rainwater Harvesting Systems." *Environmental Science & Technology* 48 (7): 4069. https://doi.org/10.1021/es500189f.

Ghimire, Santosh R., John M. Johnston, Wesley W. Ingwersen, Sarah Sojka, Wesley W. Ingwersen, and Sarah Sojka. 2017. "Life Cycle Assessment of a Commercial Rainwater Harvesting System Compared with a Municipal Water Supply System." Journal of Cleaner Production 151 (May): 74–86. https://doi.org/10.1016/j.jclepro.2017.02.025.

Godskesen, B., K. C. Zambrano, A. Trautner, N. B. Johansen, L. Thiesson, L. Andersen, and J. Clauson-Kaas, et al. 2011. "Life Cycle Assessment of Three Water Systems in Copenhagen-a Management Tool of the Future." *Water Science and Technology* 63 (3) : 565–572. doi:10.2166/wst.2011.258.

Godskesen, B., M. Hauschild, M. Rygaard, K. Zambrano, and H. J. Albrechtsen. 2013. "Life-Cycle and Freshwater Withdrawal Impact Assessment of Water Supply Technologies." *Water Research* 47 (7): 2363–2374. https://doi.org/10.1016/J.WATRES.2013.02.005.

Hofman-Caris, Roberta, Cheryl Bertelkamp, Luuk de Waal, Tessa van den Brand, Jan Hofman, René van der Aa, and Jan Peter van der Hoek. 2019. "Rainwater Harvesting for Drinking Water Production: A Sustainable and Cost-Effective Solution in The Netherlands?" *Water* 11 (3): 511. https://doi.org/10.3390/w11030511.

ISO. 2006a. "ISO 14040:2006 - Environmental Management - Life Cycle Assessment - Principles and Framework." International Organisation for Standardisation (ISO). https://www.iso.org/standard/37456.html.

ISO. 2006b. "ISO 14044:2006 Environmental Management - Life Cycle Assessment - Requirements and Guidelines." International Organisation for Standardisation (ISO). https://www.iso.org/standard/38498.html.

Jeong, Hyunju, Elizabeth Minne, and John C. Crittenden. 2015. "Life Cycle Assessment of the City of Atlanta, Georgia's Centralized Water System." *International Journal of Life Cycle Assessment* 20 (6): 880–891. https://doi.org/10.1007/S11367-015-0874-Y.

Klinglmair, Manfred, Serenella Sala, and Miguel Brandão. 2014. "Assessing Resource Depletion in LCA: A Review of Methods and Methodological Issues." *International Journal of Life Cycle Assessment* 19 (3): 580–592. https://doi.org/10.1007/s11367-013-0650-9.

Lee, Mengshan, Arturo A. Keller, Pen-Chi Chiang, Walter Den, Hongtao Wang, Chia-Hung Hou, Jiang Wu, Xin Wang, and Jinyue Yan. 2017. "Water-Energy Nexus for Urban Water Systems: A Comparative Review on Energy Intensity and Environmental Impacts in Relation to Global Water Risks." Applied Energy 205 (November): 589–601. https://doi.org/10.1016/J.APENERGY.2017.08.002.

Leong, Janet Yip Cheng, Poovarasi Balan, Meng Nan Chong, and Phaik Eong Poh. 2019. "Life-Cycle Assessment and Life-Cycle Cost Analysis of Decentralised Rainwater Harvesting, Greywater Recycling and Hybrid Rainwater-Greywater Systems." Journal of Cleaner Production 229 (August): 1211–1224. https://doi.org/10.1016/j.jclepro.2019.05.046.

Lin, Xiaohu, Wei Cheng, Juwen Huang, Guangming Li, Junlian Qiao, Jie Ren, Jingcheng Xu, et al. 2018. "Prediction of Life Cycle Carbon Emissions of Sponge City Projects: A Case Study in Shanghai, China." *Sustainability* 10 (11): 3978. https://doi.org/10.3390/su10113978.

Loubet, Philippe, Philippe Roux, Eleonore Loiseau, and Veronique Bellon-Maurel. 2014. "Life Cycle Assessments of Urban Water Systems: A Comparative Analysis of Selected Peer-Reviewed Literature." *Water Research* 67 (December): 187–202. https://doi. org/10.1016/J.WATRES.2014.08.048.

Lundie, Sven, Gregory M. Peters, and Paul C. Beavis. 2004. "Life Cycle Assessment for Sustainable Metropolitan Water Systems Planning." *Environmental Science and Technology* 38: 3465–3473. https://doi.org/10.1021/es034206m.

Mays, Larry, George P. Antoniou, and Andreas N. Angelakis. 2013. "History of Water Cisterns: Legacies and Lessons." *Water* 5 (4): 1916–1940. https://doi.org/10.3390/W5041916.

Prouty, Christine, and Qiong Zhang. 2016. "How Do People's Perceptions of Water Quality Influence the Life Cycle Environmental Impacts of Drinking Water in Uganda?" Resources, Conservation and Recycling 109 (May): 24–33. https://doi.org/10.1016/j. resconrec.2016.01.019.

Sanjuan-Delmas, David, Pere Llorach-Massana, Ana Nadal, Mireia Ercilla-Montserrat, Pere Munoz, Juan Ignacio Montero, Alejandro Josa, et al. 2018. "Environmental Assessment of an Integrated Rooftop Greenhouse for Food Production in Cities." *Journal of Cleaner Production* 177 (March): 326–337. https://doi.org/10.1016/j.jclepro.2017.12.147.

Scholz, Miklas, and Piotr Grabowiecki. 2007. "Review of Permeable Pavement Systems." *Building and Environment* 42 (11): 3830–3836. https://doi.org/10.1016/J.BUILDENV. 2006.11.016.

Slagstad, Helene, and Helge Brattebø. 2013. "Life Cycle Assessment of the Water and Wastewater System in Trondheim, Norway - A Case Study." *Urban Water Journal* 11 (4): 323–334. https://doi.org/10.1080/1573062X.2013.795232.

Sojka, Sarah, Tamim Younos, and David Crawford. 2016. "Modern Urban Rainwater Harvesting Systems: Design, Case Studies, and Impacts." In *Sustainable Water Management in Urban Environments. The Handbook of Environmental Chemistry*, edited by Tamim Younos and Tammy E. Parece, 47:209–234. Cham, Switzerland: Springer. https://doi. org/10.1007/978-3-319-29337-0_7.

Tarpani, Raphael Ricardo Zepon, Flávio Rubens Lapolli, María Ángeles Lobo Recio, and Alejandro Gallego-Schmid. 2021. "Comparative Life Cycle Assessment of Three Alternative Techniques for Increasing Potable Water Supply in Cities in the Global South." Journal of Cleaner Production 290 (March): 125871. https://doi.org/10.1016/j. jclepro.2021.125871.

Teston, Andrea, Matheus Soares Geraldi, Barbara Müller Colasio, and Enedir Ghisi. 2018. "Rainwater Harvesting in Buildings in Brazil: A Literature Review." *Water* 10 (4): 471. https://doi.org/10.3390/w10040471.

Valdez, M Carmen, Ilan Adler, Mark Barrett, Ricardo Ochoa, and Angel Perez. 2016. "The Water-Energy-Carbon Nexus: Optimising Rainwater Harvesting in Mexico City." *Environmental Processes - an International Journal* 3 (2): 307–323. https://doi. org/10.1007/s40710-016-0138-2.

Vargas-Parra, M Violeta, M Rosa Rovira-Val, Xavier Gabarrell, Gara Villalba, M Violeta Vargas-Parra, M Rosa Rovira-Val, Xavier Gabarrell, et al. 2019. "Rainwater Harvesting Systems Reduce Detergent Use." *The International Journal of Life Cycle Assessment* 24 (5): 809–823. https://doi.org/10.1007/s11367-018-1535-8.

Vieira, Abel S, Cara D Beal, Enedir Ghisi, and Rodney A Stewart. 2014. "Energy Intensity of Rainwater Harvesting Systems: A Review." *Renewable and Sustainable Energy Reviews* 34 (June): 225–242. https://doi.org/10.1016/j.rser.2014.03.012.

Wakeel, Muhammad, Bin Chen, Tasawar Hayat, Ahmed Alsaedi, and Bashir Ahmad. 2016. "Energy Consumption for Water Use Cycles in Different Countries: A Review." *Applied Energy* 178 (September): 868–885. https://doi.org/10.1016/J.APENERGY.2016.06.114.

Xue, Xiaobo, Sarah Cashman, Anthony Gaglione, Janet Mosley, Lori Weiss, Xin Cissy Ma, Jennifer Cashdollar, and Jay Garland. 2019. "Holistic Analysis of Urban Water Systems in the Greater Cincinnati Region: (1) Life Cycle Assessment and Cost Implications." Water Research X 2 (February): 100015. https://doi.org/10.1016/J.WROA.2018.100015.

Yan, Xiaoyu, Bébhinn Daly, David Butler, and Sarah Ward. 2018. "Performance Assessment and Life Cycle Analysis of Potable Water Production from Harvested Rainwater by a Decentralized System." *Journal of Cleaner Production* 172 (January): 2167–2173. https://doi.org/https://dx. doi.org/10.1016/j.jclepro.2017.11.198.

Zappone, M., S. Fiore, G. Genon, G. Venkatesh, H. Brattebø, and L. Meucci. 2014. "Life Cycle Energy and GHG Emission within the Turin Metropolitan Area Urban Water Cycle." *Procedia Engineering* 89 (January): 1382–1389. https://doi.org/10.1016/J. PROENG.2014.11.463.

2 Decentralized Urban Rainwater Harvesting in the Semiarid
A Life Cycle Assessment Based on Eco-design

Adriano Souza Leão, Thiago Barbosa de Jesus,
Hamilton de Araújo Silva Neto,
Jálvaro Santana da Hora, Samuel Alex Sipert,
Edna dos Santos Almeida, and
Eduardo Borges Cohim

2.1 INTRODUCTION

The increase in water use for anthropic activities in contrast to its availability in both quantity and quality presents challenges to urban water planning and management (Richter et al. 2018). Water demand may rise even further not only to supply urbanization and population growth but also to meet economic development trends (Brown, Keath, and Wong 2009; McDonald et al. 2014; Feng et al. 2017). In such a scenario, the capacity of centralized urban water systems (UWS) poses a risk in meeting the growing demand (Brown, Keath, and Wong 2009; McDonald et al. 2014; Feng et al. 2017).

Decentralized water sources have shown the potential to contribute to meeting the growing demand over the past decades (Schoen and Garland 2017). In regions with intense water scarcity, they aid in reductions in freshwater withdrawal from the surface and ground bodies. Rainwater harvesting systems (RWHS) are a multipurpose alternative to supply water for non-potable purposes – or potable use with further treatment. In some cases, it is the only or one of the few options for domestic water access (Araujo et al. 2021). Despite providing a non-continuous supply contingent on rainfall regimes, RWHS has shown potential for reducing the impacts of domestic water supply on the environment, human health, stormwater runoff, and UWS (Ghimire et al. 2014).

Literature on RWHS explored the potential for water-saving (Ghisi, Bressan, and Martini 2007), optimal tank design (Chiu, Liaw, and Chen 2009; Campisano and Modica 2012), economic feasibility (Anand and Apul 2011; Stec and Zeleňáková 2019), hydrologic impacts (Gires et al. 2015), stormwater runoff minimization (Araujo et al. 2021), and water quality and health risks (da Hora et al. 2017).

DOI: 10.1201/9781032638102-3

The relationship between energy use and greenhouse gas (GHG) emissions (Voinov and Cardwell 2009; Ward, Butler, and Memon 2012), as well as operational energy as an energy-intensity indicator (Chiu, Liaw, and Chen 2009; Vieira and Ghisi 2016), have been studied. However, deeper life cycle assessment (LCA) on end-of-life management and its implications on energy balance and carbon footprint is lacking (Crettaz et al. 1999; Marinoski and Ghisi 2011; Ghimire et al. 2014; Jesus, Kiperstok, and Cohim 2019; Valdez et al. 2016).

The choice of component materials, power supply, and end-of-life management dictates the RWHS' environmental and energy performance. It is still unclear which set of choices is more environmentally sustainable considering the entire life cycle from an eco-design viewpoint and whether such RWHS outperforms the conventional UWS (Silva et al. 2023), especially where rainfall is limited or not abundant, such as in semiarid regions. These insights aid the decision on whether to harvest rainwater, thus requiring further investigation.

Life Cycle Assessment is a widely used technique to estimate and compare the environmental and energy performance throughout the life of products and engineering systems (European Commission 2010). Eco-design and design for the environment (DfE) approaches largely rely on LCA to systematically ponder the entire life cycle of product systems in the design phase. It is aimed at improving environmental performance with the least negative effect on techno-economic aspects, focusing on the minimization of raw materials and waste, prioritizing renewable resources, process ecoefficiency, and resource recovery (Graedel 1998).

This study aimed to investigate the environmental and energy life cycle performance of different assemblies of decentralized urban Rainwater harvesting systems for the supply of $1.0\,m^3$ of water for non-potable purposes to a single-family house located in Brazil's semi-arid region. It focuses on well-established options for component materials, power sources, and end-of-life measures based on eco-design principles.

2.2 METHODS

The attributional process-based LCA method was employed based on ISO 14040 (ISO 2006a) and ISO 14044 (ISO 2006b). The four-step iterative framework was followed as described in the steps shown in Figure 2.1.

2.2.1 SCOPE

This study examined representative alternatives on the market for each material and energy component of the system. Two scenarios from cradle to grave were compared:

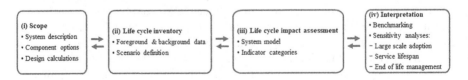

FIGURE 2.1 Methodological life cycle framework and subsections of this study.

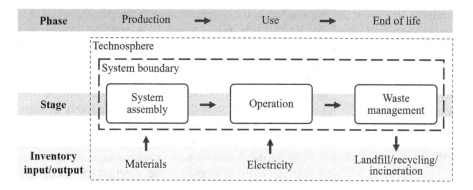

| Phase | Production | → | Use | → | End of life |

FIGURE 2.2 Life cycle boundaries of the rainwater harvesting system.

a best-case (or best assembly) with the set of lowest impact choices and a worst-case with the highest impact choices. The large-scale adoption of the best assembly in partially meeting the city's domestic demand for non-potable water, which alleviates the UWS, was evaluated. Sensitivity analyses on the service life of the system and waste management measures and assessment approaches were conducted. The system boundary comprised three phases: production (manufacture of component materials), use (operation stage), and end of life (waste management) (Figure 2.2). The target audience of this study is urban water planners, managers, and policymakers, as well as users, designers, and manufacturers.

2.2.2 System Description

The function, functional unit, and reference flow of the product system consist of the supply of $1.0\,m^3$ of rainwater harvested on-site with minimal but optimized design for non-potable uses in a low-standard single-floor house located in the 550,000 inhabitant city of Feira de Santana, State of Bahia, Brazil, with on average four permanent residents per residential unit (IBGE 2010).

The case study city is located in the northeast of the country within a semiarid region according to the Köppen-Geiger climate classification (Peel, Finlayson, and McMahon 2007), with average precipitation in the last two decades of 664 mm/year (Ghimire et al. 2014). This region is characterized by high temperatures, high evapotranspiration, and irregular precipitation (Santo and Carelli 2016), and climate and hydrological simulations for the several upcoming decades indicate a further decrease in precipitation and an increase in temperature (Tanajura, Genz, and Araújo 2010). The watershed that supplies drinking water to the city recently experienced a major drought (Oliveira 2021; ANA 2021) and has twice the water scarcity factor compared to the country-average and more than four times that of the world (Leão et al. 2022; WULCA 2016). Also, the average flow of the primary surface water source for domestic water supply is expected to decrease by 40% in the following 50 years (Valério 2015) and more than 70% by the end of the century (Genz 2011).

The safe domestic uses of roof-harvested rainwater remain under discussion. Various studies report the presence of microorganisms in rainwater tanks, but rare

are those that show the relationship of these organisms with the occurrence of diseases (Sánchez, Cohim, and Kalid 2015). While research on quantitative microbial risk assessment suggests that such a source does not meet the restrictive level of acceptability from the World Health Organization of 10^{-4} cases of disease per person per year in terms of ingestion (da Hora et al. 2017), epidemiological studies claim that drinking rainwater does not pose additional risk to illness, even when compared with water from the UWS (Heyworth et al. 2006; Rodrigo, Ledar, Sinclair 2009). Directly drinking tap water (from UWS) without further treatment on-site, such as using purifiers, is not a common practice in Brazil; therefore, drinking the water supplied by RWHS was not considered one of the domestic uses of such water in this study. A cartridge filter before the kitchen tap was adopted in the standard RWHS design, although only dish and food washing were included as on-site domestic water uses.

The adopted baseline service lifespan of the RWHS was 25 years. All the components were assumed to have a service life of 50 years, except the pump with a 15-year lifespan based on consulted literature (Parkes et al. 2010; Ghimire et al. 2014, 2017) and the cartridge refill of the kitchen tap filter every six months, fulfilling both the maximum filtered volume and the minimum replacement period requirements based on the manufacturers' data sheets (Springway 2020; Aquabios 2020; Fiel 2020). In terms of LCA modelling, replacements for pumps and cartridge refills were added at the end of their service cycles. The inventory flows and, consequently, the life cycle burdens were divided by the total water yield in the 25 years of the system's service life.

2.2.3 COMPONENT MATERIALS AND POWER SUPPLY OPTIONS

Different assemblies of RWHS consisting of representative components available on the market known to have the equivalent technical performance and capable of supplying the same yield of rainwater were considered. Figure 2.3 shows the system components for water collection (gutters), solid separation (leaf filter), transport (pipe and plumbing), storage (ground tank/cistern), pumping (pump and electricity use), a second storage before use (small header tank), and filtering for specific uses (cartridge filter before the kitchen tap). The so-called "header tank system" type considered here can be used for any pressure condition and reduces the number of pump drives per day, saving electric power.

Polyvinyl chloride (PVC) was selected as the material for pipes, plumbing, and leaf filters due to its technical performance, cost-benefit ratio, and long service life. It has been extensively used in Brazil and worldwide for these applications. The cartridge filter consisted of a polypropylene housing with activated carbon as the filter medium. Gutters of PVC, galvanized steel, and aluminium were considered. The header tank material was high-density polyethylene (HDPE). Four types of ground tanks were examined: reinforced concrete and brick as constructed on-site with an acrylic waterproof coating on the inner surface, as well as HDPE and fiberglass as prefabricated options. Of note, the ferro-cement tank commonly applied in rural RWHS was not included in this analysis, which is urban-focused. Excavation work using a hydraulic digger was considered for the ground tank. The materials of the main inner pump components considered here are cast iron and steel. The power grid

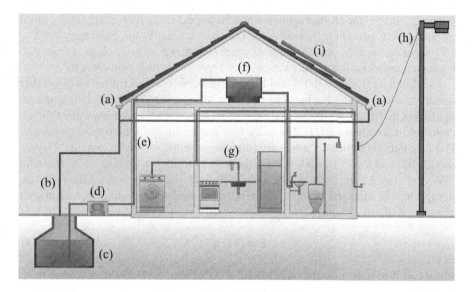

FIGURE 2.3 Schematic of the domestic rainwater harvesting system components: (a) Gutters and leaf filter, (b) pipes, (c) ground tank, (d) pump, (e) plumbing, (f) header tank, (g) cartridge filter, (h) power grid and (i) photovoltaic panels.

and on-site photovoltaic panels were adopted for electricity supply. The photovoltaic system typically adopted for on-site generation in Brazil does not include battery storage and instead delivers its converted alternating current back to the power grid.

2.2.4 OPTIMAL TANK CAPACITY AND PUMP POWER USE

The average per capita water demand in Feira de Santana of 120.0 L/(person·day) (Guanais 2015) and the average per capita roof coverage area in the northeastern region of Brazil of 19.6 m² were adopted (Ghisi 2006), thus the average roof catchment area for the 4-resident standard house modelled herein was 80.0 m².

The chosen approach to optimize the tank storage capacity by Campisano and Modica (2012) helps to balance the demand and the potential supply based on the daily rainfall data series, which is broadly deemed an effective cost-benefit procedure. The water balance in the tank was calculated using Eqs. (2.1–2.3) (Fewkes 2000).

$$Q_{(t)} = P_{(t)} C A \qquad (2.1)$$

$$Y_{(t)} = \min \begin{cases} D_{(t)} \\ V_{(t-1)} + \theta Q_{(t)} \end{cases} \qquad (2.2)$$

$$V_{(t)} = \min \begin{cases} \left(V_{(t-1)} + Q_{(t)} - \theta Y_{(t)} \right) - \left(1 - \theta \right) Y_{(t)} \\ R - \left(1 - \theta \right) Y_{(t)} \end{cases} \qquad (2.3)$$

Where $Q_{(t)}$ (m³) is the production of water with respect to the time interval t; $P_{(t)}$ (mm) precipitation; C runoff coefficient; A (m²) catchment area; $Y_{(t)}$ (m³) storage yield; $D_{(t)}$ (m³) water demand; $V_{(t)}$ (m³) volume of water stored; θ parameter ranging from 0 to 1, adopted as 0.6 for this header tank system type; and R (m³) tank capacity.

Rainfall data from 01/01/1998 to 12/31/2012 were collected from a local weather station provided by the National Institute for Meteorology (INMET 2018), and no significant difference in the annual rainfall budget was observed until 2019 (Silva, Pinto, and Castelhano 2022). No water loss during the operation was considered. The calculations indicated a 970 m³ cumulative rainwater yield over the 25-year service life and a 1,750 L tank to have the best yield. A header tank of 250 L and a 75W centrifugal pump with a 40% efficiency were sufficient to fulfil intermediate storage in this minimal design. The electricity used for pumping was estimated using Eq. (2.4).

$$E = P \cdot t \cdot \eta \tag{2.4}$$

where E (kWh) is the electric energy used; P (W), pump power; t (h), operating time; and η (%), efficiency. The calculations showed a cumulative pumping electricity of 36 kWh over 25 years.

2.2.5 Life Cycle Inventory

Table 2.1 shows the foreground inventory for both the absolute yield of the entire service life and normalized for the functional unit, consisting of primary data from the system design parameters, the National System for Research of Costs and Indices of

---

TABLE 2.1

Inventory of the Rainwater Harvesting System from Cradle-to-Grave for an Absolute Yield of 970 m³ of Water in 25 Years and Normalized for the Functional Unit of 1.0 m³ of Water Supplied

		Value for			
Item	Component Options	25-Year Service Life (970 m³)	Functional Unit (1.0 m³)	Unit	Source
		Inputs			
Ground tank	1. Concrete	431	4.44×10^{-1}	kg	Eq. (2.3); [1]
		1,085	$1.12 \cdot \times 10^{0}$	kg	
		216	$2.23 \cdot \times 10^{-1}$	kg	
		1,506	$1.55 \cdot \times 10^{0}$	kg	
		128	$1.32 \cdot \times 10^{-1}$	kg	
		11	$1.13 \cdot \times 10^{-2}$	kg	
		5	$5.15 \cdot \times 10^{-3}$	kg	

(Continued)

TABLE 2.1 (*Continued*)

Inventory of the Rainwater Harvesting System from Cradle-to-Grave for an Absolute Yield of 970 m³ of Water in 25 Years and Normalized for the Functional Unit of 1.0 m³ of Water Supplied

Item	Component Options	Value for 25-Year Service Life (970 m³)	Value for Functional Unit (1.0 m³)	Unit	Source
	2. Brick	218	$2.25 \cdot \times 10^{-1}$	kg	Eq. (2.3); [1]
		195	$2.01 \cdot \times 10^{-1}$	kg	
		1,148	$1.18 \cdot \times 10^{0}$	kg	
		392	$4.04 \cdot \times 10^{-1}$	kg	
		454	$4.68 \cdot \times 10^{-1}$	kg	
		55	$5.67 \cdot \times 10^{-2}$	kg	
		10	$1.03 \cdot \times 10^{-2}$	kg	
		5	$5.15 \cdot \times 10^{-3}$	kg	
	3. HDPE	34	$3.51 \cdot \times 10^{-2}$	kg	Eq. (2.3); [2]
	4. Fiberglass	31	$3.20 \cdot \times 10^{-2}$	kg	Eq. (2.3); [3]
		10	$1.03 \cdot \times 10^{-2}$	kg	
	Excavation	6	$6.19 \cdot \times 10^{-3}$	m³	[1]
Gutters	1. PVC	15	$1.55 \cdot \times 10^{-2}$	kg	[4]
	2. Galvanized steel	18	$1.86 \cdot \times 10^{-2}$	kg	[5]
	3. Aluminium	9	$9.28 \cdot \times 10^{-3}$	kg	[6]
Pipe/plumbing	PVC	23	$2.37 \cdot \times 10^{-2}$	kg	Estimated
Leaf filter	PVC	1	$1.03 \cdot \times 10^{-3}$	kg	Estimated
Cartridge filter	Housing	1	$1.03 \cdot \times 10^{-3}$	kg	[7]
	Cartridge refill	5	$5.15 \cdot \times 10^{-3}$	kg	
Header tank	HDPE	7	$7.22 \cdot \times 10^{-3}$	kg	[2]
Pump	Steel and others	2	$2.06 \cdot \times 10^{-3}$	p	-
Electricity	1. Power grid	36	$3.71 \cdot \times 10^{-2}$	kWh	Eq. 4
	2. Photovoltaic	36	$3.71 \cdot \times 10^{-2}$	kWh	
	Outputs				
Water supply	Reference product	970	1.0	m³	Eq. 2
Waste management	1. Landfill	*	*	kg	Estimated
	2. Recycling				
	3. Energy recovery				

*End-of-life flow values vary by scenario.
[1] SINAPI CEF (2019); [2] Fortlev (2019a); [3] Fortlev (2019b); [4] Tigre (2019); [5] Mopa (2019); [6] Shockmetais (2019); [7] Pentair (2014).

Civil Construction from Caixa Econômica Federal (SINAPI CEF 2019), and manufacturer data sheets. The assemblies consisted of one type of material for each of the listed component options. Inflows are the component materials and power supply.

Outflows are the water supply (reference product) and waste management measures: landfill disposal, recycling, or incineration. An equivalent availability of components for retail within the city and minor distances for construction and demolition logistics were assumed; hence, transport of components was disregarded.

2.2.6 Best-case and Worst-case Scenarios

Two scenarios were proposed and evaluated:

- Best-case: assembly consisting of the system components with the highest environmental and energy performance based on eco-design principles.
- Worst-case: assembly consisting of the system components with the lowest environmental and energy performance for reference.

Both scenarios included some of the same components, so-called herein 'fixed components' namely: pipes, plumbing, leaf filter, cartridge filter, header tank, pump, and excavation works for the ground tank. The technical performance of the RWHS was assumed to be the same; therefore, the functional unit of the product system remained unchanged. In addition, landfill disposal was assigned as the baseline option for waste management. Only the photovoltaic system was not fully accounted for final disposal; 10% of its bulk mass was allocated to end-of-life management due to the RWHS, since it indirectly offsets the power supply to other home appliances during its service life by supplying the power grid.

2.2.7 Life Cycle Impact Assessment

The SimaPro 9.1 software was used with the Ecoinvent 3.5 cut-off database for the background inventories considering Brazil {BR} or the rest of the world {RoW} regions. Carbon and energy footprints were the midpoint categories assessed using the Global Warming Potential (GWP) IPCC 100 years and the Cumulative Energy Demand (CED) – including renewable and non-renewable subcategories – characterization methods. The GWP indicator accounts for GHG atmospheric emissions along the life cycle of a product, activity, or service (Frischknecht et al. 2007). Different contributions to global warming are weighed and converted into CO_2 equivalents, being released directly by a process or indirectly by the power supply, for instance. CED is a technical category that has been used as a proxy for environmental issues (Oregi et al. 2015). The CED indicator quantifies the energy content of several renewable and non-renewable energy resources (Frischknecht et al. 2007). For instance, the production of energy-carrying materials and electricity requires fuels and energy to manufacture components and run machines. Additionally, the use of raw materials is typically associated with energy demand indicators.

2.2.8 Scale-up Adoption

To grasp whether the adoption of the best-case RWHS in a decentralized manner could be beneficial to the environmental performance of the city's water domestic supply, a scale-up analysis based on the adoption of RWHS throughout the city's domestic built environment was conducted. This comparison was based on the environmental performance indicators of the UWS of the same municipality (Guanais, Cohim, and Medeiros 2017; Guanais 2015). The following premises and data were considered:

- The net volume of drinking water (after losses) consumed in Feira de Santana in 2017 equals $22{,}530{,}300\,m^3$ (EMBASA 2018);
- The same per capita consumption adopted in the previous section of 120.0 L/(person·day), resulting in a demand of about $175.2\,m^3$ per year for the standard house considered here;
- A rainwater yield of $38.8\,m^3$ per year for non-potable purposes (Eq. 2.3);
- The demand and supply assumptions above result in a number of single-family houses of around 130 000, which is in the same order of magnitude as the surveyed number of houses or domestic units in the city (IBGE 2010).

In this way, the life cycle environmental and energy performance were estimated for the city's water demand for domestic use in a year as a function of the ratio (%) of hypothetical standard houses that would adopt such RWHS.

2.2.9 SERVICE LIFE SENSITIVITY ANALYSIS

The Brazilian standard for the performance of residential buildings (ABNT 2013) establishes a minimum service life of 20 years for hydro-sanitary systems considering appropriate maintenance, 25 years for the "intermediate" span, and 30 years for the "superior" specification. The Brazilian standard for rainwater harvesting (ABNT 2019) indicates a maintenance check schedule of a year for the tank; a semester for gutters, pipes, and plumbing; and a month for filters and pumps. A longer service life of the components with minor maintenance demands is expected when following this check schedule, as long as the system remains undamaged by movement or impacts. For instance, tanks can last for over 100 years under appropriate conditions (Parkes et al. 2010).

While related literature has considered a service life of 50 years (Ghimire et al. 2014), and although the materials may not critically degrade along this timespan in indoor and well-maintained conditions, the components can lose functionality and require significant maintenance or repair over several decades of use (Parkes et al. 2010). Additionally, the longer the service life, the more water is provided by the system, which "dilutes" the contribution of the components to the overall life cycle burdens within the approach adopted in this study. Therefore, the intermediate value of 25 years of the performance standard (ABNT 2013) was adopted as the baseline service life, which is not only a conservative option but also deemed to be representative of both the integrity of the components and a reasonable timespan for potentially significant intervention.

Considering the abovementioned, a sensitivity analysis of extended service lives was carried out for 30 years, which is aligned with the upper bound performance of hydro-sanitary systems (ABNT 2013), 50 years as adopted in similar studies (Ghimire et al. 2014, 2017), and 40 years, which is in between.

2.2.10 END-OF-LIFE SENSITIVITY ANALYSIS

Sensitivity analyses (i, ii, iii) of the end of life of the best-case scenario were conducted to investigate waste management options other than landfill disposal and different LCA assessment approaches.

i. Full recycling was assumed for the wastes from PVC pipes, plumbing and
    leaf filters, polypropylene filter housing, HDPE tanks, metal pumps, gal-
    vanized steel gutters, and part of the photovoltaic system. The recycling of
    waste materials under the cut-off approach implies disposing of less solid
    waste in landfills, therefore offsetting burdens by transferring them to new
    consumer product systems.
ii. Energy recovery via incineration of plastic wastes was modelled as fur-
    ther waste treatment, considering both positive and negative outcomes from
    an LCA viewpoint alternatively to the cut-off approach. Up to 71% of the
    embodied energy in HDPE could be deemed as feedstock energy, and simi-
    larly, 36% in PVC and 56% in polypropylene (Hammond, G., Jones 2008).
    An incinerator with a 25% energy recovery efficiency was considered
    (Wernet et al. 2016; Moreno Ruiz et al. 2018).
iii. A system expansion by substitution was carried out, which comprised not
    only the pros and cons of plastic waste incineration with energy recovery
    but also the effects of the substitution of a conventional fuel used in energy
    production. For this, it was considered a balance between the burdens of
    plastic waste treatment and the avoided burdens of heavy oil or natural gas
    burning to produce the same amount of energy, as well as the burdens of the
    production of the fuel itself.

2.3 RESULTS AND DISCUSSION

The environmental and energy performance of each system component, followed by
the best-case and worst-case assemblies, was presented. The sensitivity and benefits
of RWHS adoption on a city scale, extended service life, as well as end-of-life management and modelling, were discussed.

2.3.1 ENVIRONMENTAL AND ENERGY PERFORMANCE OF
THE SYSTEM COMPONENTS

The results show how each choice affects the overall RWHS environmental burdens
(Figure 2.4, Table 2.2).

Piping and plumbing had the highest impact among the fixed components, and
the excavation of the ground tank had the lowest contribution. For GWP, the filters
yielded the second largest burden, followed by the pumps and header tank. However,
the pumps were less energy-demanding than the HDPE header tank, whose energy
sources were 93% fossil-based resources. As for the fraction of non-/renewable
resources in the embodied energy, all the fixed components consisted mostly of
non-renewable resources, whose content was at least 80% fossil.

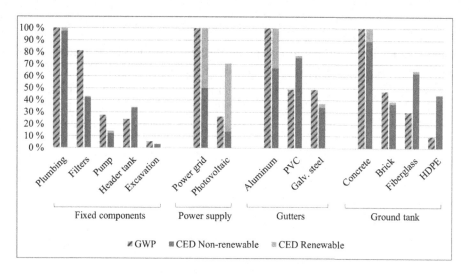

FIGURE 2.4 Relative life cycle environmental and energy performance of alternatives to system components.

TABLE 2.2
Life Cycle Environmental and Energy Performance of the RWHS Components and Scenarios per Functional Unit (1.0 m³ of Water Supplied)

Item	GWP (kg CO_2-eq·m^{-3})	CED (MJ·m^{-3})	Non-renewable CED (MJ·m^{-3})	Renewable CED (MJ·m^{-3})
		Components		
Production phase **Fixed components**				
Pipes and plumbing	6.00×10^{-2}	1.63	1.59	3.94×10^{-2}
Filters	4.87×10^{-2}	7.05×10^{-1}	6.89×10^{-1}	1.56×10^{-2}
Pump	1.65×10^{-2}	2.35×10^{-1}	2.02×10^{-1}	3.35×10^{-2}
Header tank	1.45×10^{-2}	5.58×10^{-1}	5.51×10^{-1}	6.50×10^{-3}
Excavation works	3.32×10^{-3}	5.15×10^{-2}	5.11×10^{-2}	3.71×10^{-4}
Electricity source				
Power grid	1.05×10^{-2}	2.63×10^{-1}	1.32×10^{-1}	1.31×10^{-1}
Photovoltaic	2.77×10^{-3}	1.86×10^{-1}	3.70×10^{-2}	1.49×10^{-1}
Gutter				
Aluminium	8.80×10^{-2}	1.38	9.20×10^{-1}	4.60×10^{-1}
PVC	7.82×10^{-2}	1.25	1.22	3.02×10^{-2}
Galvanized steel	4.29×10^{-2}	5.16×10^{-1}	4.73×10^{-1}	4.29×10^{-2}

(Continued)

TABLE 2.2 (*Continued*)
Life Cycle Environmental and Energy Performance of the RWHS Components and Scenarios per Functional Unit (1.0 m³ of Water Supplied)

Item		GWP (kg CO_2-eq·m^{-3})	CED (MJ·m^{-3})	Non-renewable CED (MJ·m^{-3})	Renewable CED (MJ·m^{-3})
			Components		
	Fixed components				
	Tank				
	Concrete	7.16×10^{-1}	6.05	5.41	6.39×10^{-1}
	Brick	3.39×10^{-1}	2.37	2.24	1.24×10^{-1}
	Fiberglass	2.16×10^{-1}	3.91	3.79	1.26×10^{-1}
	HDPE	7.04×10^{-2}	2.71	2.68	3.16×10^{-2}
			Scenarios		
Best-case					
Production phase/ system assembly	Fixed components	1.43×10^{-1}	3.18	6.29	3.19×10^{-1}
	HDPE tank	7.04×10^{-2}	2.71		
	Galvanized steel gutter	4.29×10^{-2}	5.16×10^{-1}		
Use phase/ operation	Photovoltaic electricity	2.77×10^{-3}	1.86×10^{-1}		
End-of-life phase	Waste disposal	1.23×10^{-2}	1.55×10^{-2}		
Worst-case					
Production phase/ system assembly	Fixed components	1.43×10^{-1}	3.18	1.06×10	1.33
	Concrete tank	7.16×10^{-1}	6.05		
	Aluminium gutter	8.80×10^{-2}	1.38		
Use phase/ operation	Power grid electricity	1.05×10^{-2}	2.63×10^{-1}		
End-of-life phase	Waste disposal	5.39×10^{-2}	1.10		

On-site photovoltaic power supply outperformed the power grid by about 74% for GWP and 29% for CED. Renewable resources represented 80% of the photovoltaic system energy demand, of which 96% was directly from the sun, while they represented around 50% of the power grid energy demand, of which 97% was from water. The power grid system comprises not only production per se, which reflects the average matrix of the country or region, but also infrastructure and losses from transmission, distribution, and voltage transformations to reach the end-user.

In the base year of the inventory dataset used (2014), the Brazilian grid consisted of 75% of renewable resources, of which 87% were from hydropower (EPE 2015), whereas in 2020, renewable power was 85% of the matrix, being 77% of that from hydropower, while non-renewable power was 77% on average worldwide (EPE 2021). Photovoltaic power generation, both decentralized (on-site) and centralized (photovoltaic "farms" or plants), has been largely adopted across the country in recent years, providing 2% of the total power supply in 2020 (EPE 2021). Photovoltaic energy losses and efficiency reductions are primarily caused by the heating of the panels, dust accumulation and direct-to-alternating current transformation to deliver to the power grid (Ferroni and Hopkirk 2016). Although both photovoltaic and hydroelectric plants are renewable and rely directly on the sun as a source, other centralized sources within the power matrix, such as thermal power plants, as well as the grid infrastructure and its inherent losses, increase the environmental burdens of the power grid supply.

Aluminium gutters presented the largest GWP and CED among the options, followed by PVC and galvanized steel, with 51% less GHG emissions than aluminium and at least 52% less energy demand than the other two options. In terms of market share and cost-benefit, PVC stands out, although galvanized steel and aluminium are also usual options. While gutters of aluminium are the lightest on a mass basis because aluminum allows for manufacturing slender components, it is an energy-intensive material with a significant indirect carbon footprint. Nevertheless, the energy demanded by the average aluminium production in South America has a significant renewable content, primarily due to the hydropower supply of Brazil's electricity matrix, resulting in a lower non-renewable CED than PVC but still larger than galvanized steel. The latter had up to 88% nonrenewable embodied energy due to the dependence on coke and hard coal in the pig iron and steel supply chain (Wernet et al. 2016; Moreno Ruiz et al. 2018).

The carbon footprint ranking for ground tanks was concrete, brick, fiberglass, and HDPE, while the brick tank presented the lowest energy footprint. Cement was the concrete raw material with the highest GWP (56%), due to its clinkerization process, which releases a great amount of CO_2 directly through the calcination of limestone and fuel burning in the kiln, while reinforcing steel was a major contributor to CED (56%). The use of steel in concrete tanks is significant due to the high reinforcement density required to support design loads and moments from hydraulic pressure and sharp corners. About 11% of the concrete tank CED was renewable, mainly due to the wood formwork and hydropower share in the power supply for cement and steel manufacture. The embodied energy of the other tank options was at least 95% non-renewable. The largest brick tank burdens come from cement (53% GWP and 33% CED), followed by clay bricks. The polyester resin was a major contributor to the fiberglass tank (>88% GWP and CED). The HDPE tank GWP comes from direct CO_2 emissions in manufacturing and crude oil for CED. In terms of material demand, reinforced concrete and brick tanks had a bulk mass between 60 and 99 times that of plastic tanks, which may also affect transport burdens.

2.3.2 ENVIRONMENTAL AND ENERGY PERFORMANCE OF BEST-CASE AND WORST-CASE SCENARIOS

Based on the component contributions and under an eco-design perspective, the chosen best-case scenario was composed of a galvanized steel gutters, an HDPE tank, and photovoltaic electricity supply; and the worst-case scenario consisted of an aluminium gutter, a concrete tank, and a power grid electricity supply, in addition to the fixed components (Figure 2.5, Table 2.2). Despite being highly renewable, the Brazilian power grid still falls behind photovoltaics in this respect for general applications; however, such a trade-off needs to be considered case by case.

The best assembly presented about 73% less GWP as well as 45% less CED comparatively, with GWP of 0.271 kg CO_2-eq/m^3 and CED of 6.61 MJ/m^3, while the worst assembly has 1.01 kg CO_2-eq/m^3 and 12.0 MJ/m^3. The HDPE tank presented lower GWP and higher CED compared with the brick tank. The HDPE tank was assigned to the best-case because of the greater relative difference in GWP compared with the concrete tank. It also has 60 times less bulk mass, resulting in lower burdens in final disposal as well as the existence of known recycling and reuse routes. Concrete tank production stood out among the worst assembly components (71% GWP and 51% CED), followed by the set fixed components, aluminium gutter, waste disposal, and power grid electricity. The set of fixed components stood out for the GWP and HDPE ground tank for CED in the best assembly. The fixed components were the second major contributor for CED and the HDPE ground tank for GWP. Galvanized steel gutters, photovoltaic power, and waste disposal were last in the rankings. Choosing low-burden components increased the proportional contribution of the fixed components to the overall indicator values substantially.

The production phase was dominant in both categories, ranging from 89% to 97%, which can be directly associated with the manufacture of the component materials (Figure 2.6). The use phase stood for only 1%–3% of GWP and CED, attributed solely to the power supply. The worst-case presented larger burdens in

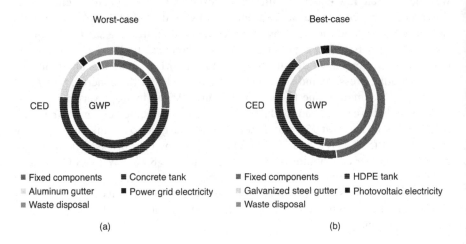

FIGURE 2.5   Contribution analysis of (a) worst-case and (b) best-case scenarios.

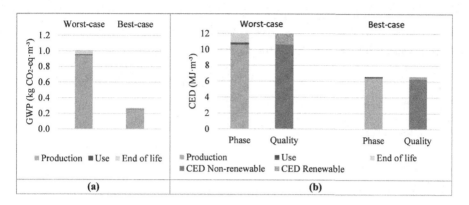

FIGURE 2.6 Dominance analysis of life cycle phases per scenario for (a) GWP and (b) CED, including share of energy non-/renewable resources (quality).

the end-of-life phase, mainly due to the heavier tank (5% GWP and 9% CED), while end of life was less representative in the best-case (3% GWP and 0.3% CED). The worst-case showed greater renewable energy content in absolute and relative terms (11% CED), while the best-case presented a lower relative dependence on non-renewables (5% CED).

2.3.3 Life Cycle Performance Benchmarking

The range of GWP and CED of the worst and best assemblies was in the same order of magnitude compared with the consulted literature (Table 2.3).

Ghimire et al. (2014) evaluated an RWHS with typical and minimal infrastructure and energy use. Their findings showed that a typical RWHS did not perform better than the UWS in all assessed categories in Roanoke, Virginia, United States, while a minimal design did. The major contributions of the typical design were attributed to the power supply for pumping directly to the point of use, followed by the HDPE tank. In the case study reported here, the best-case showed a 14% greater CED and 34% lower GWP compared with their minimal design.

In another case study, Ghimire et al. (2017) found that a benchmark commercial RWHS performed better than the UWS for GWP and CED in Washington, D.C., United States. The power supply for pumping in the treatment step and to the point of use had a major impact, followed by the fiberglass tank. In the case study reported here, the best-case performed better in the two categories. Both studies (Ghimire et al. 2014, 2017) assessed a pressurized type of system with no header tank called the "direct feed" system, which increases the impacts of the operation phase.

Angrill et al. (2012) assessed the use of rainwater to supply laundry systems in Europe. They examined a two-story detached single-family scenario and a five-story multi-apartment scenario with gravity and header tank systems, in which the latter requires pumping. The GWP of their single-family scenario with pumping was 83% attributed to the materials, majorly to the concrete tank; 10% to operation, i.e.,

TABLE 2.3

Comparison with the Results of the Consulted Literature

Case Study	GWP (kg CO_2-eq $\times m^{-3}$)	Energy Demand (MJ $\times m^{-3}$)	Reference
Domestic RWHS in Feira de Santana, Bahia, Brazil	0.271–1.01	6.61–12.0	[1]
Domestic RWHS in Roanoke, Virginia, USA	0.41	5.8	[2]
Commercial RWHS in Washington, D.C., USA	0.33	6.8	[3]
UWS of Feira de Santana, Bahia, Brazil	0.47	12.6	[4, 5]

[1] This study; [2] Ghimire et al. (2014); [3] Ghimire et al. (2017); [4] Guanais (2015); Guanais et al. (2017).

power supply, 5% to transport, and 1% to construction and deconstruction altogether. The dominance of the production phase over operation and post-use was similarly observed in the case study reported here. Consistent result comparison was not possible due to differences in the system function and functional units.

Marinoski and Ghisi (2011) studied RWHS for toilet flushing supply in a detached single-family house in southern Brazil. They conducted a sensitivity analysis on the main system components for a range of demand ratios, considering different ground tank materials. Their findings show that the greater the rainwater demand, the greater the GWP and CED due to the required increase in tank size. The concrete tank presented greater burdens than the HDPE option, followed by fiberglass. Their functional unit was not comparable to the case study reported here.

Comparing the water supply by RWHS and UWS poses consistency issues because the product system functions are not strictly equivalent, which is relevant in comparative LCA. Generally, a UWS is designed to provide potable water continuously or with reasonably controlled gaps of shortage. A standard RWHS generally supplies the water demand partially, non-continuously, and for non-potable purposes; otherwise, further treatment is required on-site. The quality of the water source also plays a major role, as rainwater usually requires less treatment effort than surface or ground freshwater. Overall, such functions have a significant overlap since domestic non-potable uses are predominant.

A UWS demands electricity in several stages, from water extraction to treatment and distribution. At the same time, power generation also relies on the potential energy harvested from water cycles, especially in the case of the Brazilian matrix. This relationship is often called the water-energy nexus. Additionally, the more energy-intensive the water supply, the greater the pressure on the environment (Hamiche, Stambouli, and Flazi 2016), even for highly renewable electricity matrices that still partially rely on intense carbon footprint power generation (Voinov and Cardwell 2009).

Concerning embodied energy, while the UWS requires a considerably larger infrastructure to be constructed and maintained compared with a single or even multiple RWHS, it also provides a substantially larger volume of water throughout its life cycle. However, its efficiency is directly affected by distribution losses. Similar to the concept of economy of scale, although the environmental impacts generated by the UWS infrastructure and maintenance could be significant in absolute terms, they might be offset by its huge water yield.

Guanais (2015) and Guanais et al. (2017) assessed the UWS in the same municipality in this case study, using a similar approach and background data. Their findings showed that 75% of GWP and 86% of CED were due to the power supply for pumping in several stages of the system. Chemicals for treatment were the second-highest contribution, followed by infrastructure maintenance. The transport of replacement piping and chemicals was negligible; thus, this comparison remains consistent. Regarding energy demand, both the worst and the best RWHS assemblies reported herein outperformed the municipal system, especially the best-case, with 48% and 42% smaller CED and GWP, respectively.

2.3.4 BENEFITS OF RWHS ADOPTION ON A CITY SCALE

Absolute GWP and CED indicators were estimated for the annual domestic water demand of the entire city as a function of the ratio (%) of hypothetic single-family standard houses modelled herein that would adopt such an RWHS (Figure 2.7).

In addition to minimizing the water demand from the UWS, an increasing adoption ratio of the RWHS in the residential built environment would reduce the carbon and energy footprints of the city's domestic water supply in absolute terms. For instance, a 25% adoption ratio could have avoided the demand for 9.8 million m^3 of freshwater, an emission of 248 t of CO_2-eq, and a demand of 7.5 TJ of energy for domestic water supply in Feira de Santana in 2017, located in a semiarid area with

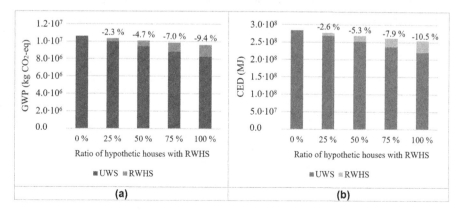

FIGURE 2.7 Environmental and energy performance of domestic water supply in a year as a function of the RWHS adoption ratio for (a) GWP and (b) CED in absolute terms.

limited rainfall. This is reassuring and should further support policy and policymakers' efforts to encourage large-scale RWHS deployment, for instance, by providing a discount on the annual municipal property tax.

Such benefits could help achieve large-scale environmental gains and targets. In the Paris Climate Agreement, Brazil pledged to decrease GHG emissions per capita per year from 6.5 t in 2012 to 6.2 t in 2025, considering population growth (Brazil 2015). The GHG emission per capita in the state of Bahia was 6.3 t in 2017 (SEEG 2020). Considering 25% of the city's population, the avoided emissions in domestic water supply would be equivalent to about 0.6% of the national goal and 1.5% for a Bahia state citizen, and the energy-savings would be equivalent to about 0.4% of the domestic electricity use in Feira de Santana in 2016 based on data from the Municipal Information System (SIM 2017).

2.3.5 Extended-Service Life Benefits

Extended service life is one of the premises of eco-design. A longer lifespan for RWHS implies lower burdens per m³ supplied (Figure 2.8). From 25 to 30 years, a reduction of 17% in both GWP and CED was observed; similarly, going from 25 to 40 years resulted in a 37% lower GWP and CED; and from 25 to 50 years GWP decreased by 39% and CED by 44%. A 25-year service life as a baseline was a conservative choice; hence, the comparison of the RWHS with the UWS and the scaling-up of its adoption across the city would lead to even better outcomes for longer service lives.

Using linear regression to model the decrease in burdens for increasing service life, the coefficient of determination R^2 was approximately 1.0 up to the fourth decimal figure for a fitted polynomial curve of order 2 for both the GWP and CED, fitting better than the linear, exponential, logarithmic, and power curves. This non-linearity is due to the pump and the cartridge refill replacements over the RWHS modelled lifespan, which affected both the production and end-of-life phases. Otherwise, a linear trend would be expected, which is the baseline assumption of the attributional process (LCA).

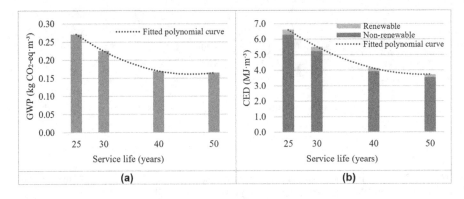

FIGURE 2.8 Sensitivity analysis of the best-case assembly's extended service life for (a) GWP and (b) CED.

2.3.6   Effect of End-of-life Management and Modelling

In a comprehensive eco-design analysis of the environmental aspects and impacts of engineering systems, the decommissioning stage is a key concern, and different measures and modelling assumptions based on the best-case scenario led to divergent outcomes (Figure 2.9).

i. Even though the contribution of landfill disposal in the best-case scenario was not a major impact, recycling could minimize the GWP by around 4.4% and the CED by 0.2%.

ii. Energy recovery from plastic wastes through incineration could reduce CED by 12.1%. In contrast, since HDPE, PVC, and polypropylene are composed mostly of fossil hydrocarbons, burning tanks, pipes, plumbing, filters, and the plastic content of the photovoltaic system would give rise to expressive GHG emissions, increasing GWP by 64.1%.

iii. Considering plastics as an alternative energy source to partially offset the burdens of a fuel burned outside of the system boundary showed a potential reduction of 16.2% in the CED of the baseline option, but still a substantial increase of 39.1% in GWP. While the avoided burning of heavy oil or natural gas could offset the overall GHG emissions to some extent, the atmospheric emission of fossil carbon by harvesting energy from plastics is still significant. These results are aligned with the current discussion on the fate of plastic waste as an energy source in the post-use phase. While energy harvesting as an end-of-life measure permits further use of resources stored in plastics, the effects of the current approach to such a process are deemed environmentally unsustainable (McGuirk, Bazilian, and Kammen 2019).

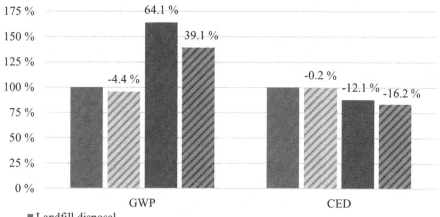

FIGURE 2.9   Sensitivity analysis of the best-case assembly's end-of-life measures and modelling approaches.

2.4 CONCLUSION

This study endorses the benefits of urban Rainwater harvesting systems with respect to their environmental and energy performance, even in a semiarid region, considering a minimal but optimized design and non-potable supply. Eco-design thinking was decisive to the life cycle performance of the RWHS.

The best-case RWHS, consisting of the most environmentally friendly components among well-established options on the market, resulted in a Global Warming Potential (GWP) of 0.271 kg CO_2-eq/m^3 and Cumulative Energy Demand (CED) of 6.61 MJ/m^3, 45%–73% lower compared with the worst-case. The production phase stood out compared with the operation and end-of-life phases, representing 89%–97% of the overall burdens. The ground tanks were the top contributors among the system components, especially the concrete tank.

The best-case RWHS outperformed the local urban water supply (UWS) by 42% in GWP and 48% in CED per m^3 supplied. Scaling up the adoption of RWHS in the city's residential built environment could significantly improve the environmental and energy performance of the domestic water supply. For instance, a 25% coverage would save nearly 9.8 million m^3/year of freshwater and 7.5 TJ/year of energy use and minimize 248 t CO_2-eq/year in GHG emissions. The service life sensitivity analysis showed that longer-lasting systems lead to lower impacts per m^3 supplied, following a nonlinear trend due to different component replacement frequencies.

The end-of-life sensitivity analysis also showed that avoiding waste disposal in landfills could marginally decrease the burdens, considering the cut-off approach for recycling. While energy recovery from plastic waste upon incineration could be energetically beneficial compared with landfill (baseline end of life management), with a decrease in CED of 12%–16% depending on the LCA modelling, it presented negative outcomes in terms of unintended emissions similar to those of energy production from conventional fossil fuels, as long as no further measures for emission control were taken, increasing GWP by 39%–64%.

Future research may investigate different design options and data pertaining to other regions and conditions, as well as LCA modelling approaches, i.e., consequential and inputoutput, in addition to the assessment of other impact categories, the combination of RWHS with water reuse and water reduction techniques, and uncertainty and cost analysis for distinct adoption scales.

REFERENCES

ABNT (Brazilian Association of Technical Standards). 2013. "NBR 15575: Residential Buildings - Performance." ABNT (Brazilian Association of Technical Standards). 2019. "NBR 15527: Rainwater Harvesting from Roofs for Non-Potable Uses - Requirements."

ANA. 2021. "Monitor de Secas [Drought Monitor Base Year 2019]." https://monitordesecas.ana.gov.br/mapa?mes=1&ano=2019.

Anand, C., and D.S. Apul. 2011. "Economic and Environmental Analysis of Standard, High Efficiency, Rainwater Flushed, and Composting Toilets." *Journal of Environmental Management* 92 (3): 419–428. https://doi.org/10.1016/j.jenvman.2010.08.005.

Angrill, Sara, Ramon Farreny, Carles M. Gasol, Xavier Gabarrell, Bernat Viñolas, Alejandro Josa, and Joan Rieradevall. 2012. "Environmental Analysis of Rainwater Harvesting

Infrastructures in Diffuse and Compact Urban Models of Mediterranean Climate." *The International Journal of Life Cycle Assessment* 17 (1): 25–42. https://doi.org/10.1007/s11367-011-0330-6.

Aquabios. 2020. "Technical Catalog of Cartridge Filters and Purifiers." https://www.filtrosacquabios.com.br/catalogo/samples/magazine/catalogo_produtos1220.pdf.

Araujo, M. C., A. S. Leão, T. J. Barsosa, and E. B. Cohim. 2021. "The Role of Rainwater Harvesting in Urban Stormwater Runoff in the Semiarid Region of Brazil." *Urban Water Journal* 18 (4) : 248–256. https://doi.org/10.1080/1573062X.2021.1877743.

Brazil. 2015. "Pretendida Contribuição Nacionalmente Determinada Para Consecução Do Objetivo Da Convenção-Quadro Das Nações Unidas Sobre Mudança Do Clima [Intended Nationally Determined Contribution to Achieve the Goal of the United Nations Convention on Climate Change]." United Nations Framework Convention on Climate Change. https://www.gov.br/mre/pt-br/arquivos/documentos/clima/brasil-indc-portugues.pdf.

Brown, R. R., N. Keath, and T. H. F. Wong. 2009. "Urban Water Management in Cities: Historical, Current and Future Regimes." *Water Science and Technology* 59 (5): 847–855. https://doi.org/10.2166/wst.2009.029.

Campisano, Alberto, and Carlo Modica. 2012. "Optimal Sizing of Storage Tanks for Domestic Rainwater Harvesting in Sicily." *Resources, Conservation and Recycling* 63 (June): 9–16. https://doi.org/10.1016/j.resconrec.2012.03.007.

Chiu, Yie-Ru, Chao-Hsien Liaw, and Liang-Ching Chen. 2009. "Optimizing Rainwater Harvesting Systems as an Innovative Approach to Saving Energy in Hilly Communities." *Renewable Energy* 34 (3): 492–498. https://doi.org/10.1016/j.renene.2008.06.016.

Crettaz, P, O Jolliet, J Cuanillon, S Orlando, Ecole Polytechnique, and Feâdeârale De Lausanne. 1999. "Life Cycle Assessment of Drinking Water and Rain Water for Toilets Flushing." *Journal of Water Supply: Research and Technology* 48 (3): 73–83.

EMBASA (Bahia Water and Sanitation Company). 2018. "Controle Operacional de Água e Esgoto Do Município de Feira de Santana Para o Ano de 2016 [Operational Control of Water and Sewage of the Municipality of Feira de Santana for the Year 2016]."

EPE. 2021. "Balanço Energético Nacional Relatório Síntese: Ano Base 2020 [National Energy Balance Summary Report: Base Year 2020]." https://www.epe.gov.br/pt/publicacoes-dados-abertos/publicacoes/balanco-energetico-nacional-2021.

EPE (Company of Energetic Research). 2015. "Balanço Energético Nacional Relatório Síntese: Ano Base 2014 [National Energy Balance Summary Report: Base Year 2014]." http://www.epe.gov.br/pt/publicacoes-dados-abertos/publicacoes/Balanco-Energetico-Nacional-2015.

European Commission. Joint Research Centre. Institute for Environment and Sustainability. 2010. "International Reference Life Cycle Data System (ILCD) Handbook - General guide for Life Cycle Assessment - Detailed guidance", 1st edn. EUR 24708, EN. Luxembourg. Publications Office of the European Union. https://eplca.jrc.ec.europa.eu/uploads/ILCD-Handbook-General-guide-for-LCA-DETAILED-GUIDANCE-12March2010-ISBN-fin-v1.0-EN.pdf.

Feng, Le, Bin Chen, Tasawar Hayat, Ahmed Alsaedi, and Bashir Ahmad. 2017. "The Driving Force of Water Footprint under the Rapid Urbanization Process: A Structural Decomposition Analysis for Zhangye City in China." *Journal of Cleaner Production* 163 (October): S322–S328. https://doi.org/10.1016/j.jclepro.2015.09.047.

Ferroni, Ferruccio, and Robert J. Hopkirk. 2016. "Energy Return on Energy Invested (ERoEI) for Photovoltaic Solar Systems in Regions of Moderate Insolation." *Energy Policy* 94 (July): 336–344. https://doi.org/10.1016/j.enpol.2016.03.034.

Fewkes, A. 2000. "Modelling the Performance of Rainwater Collection Systems: Towards a Generalised Approach." *Urban Water* 1 (4): 323–333. https://doi.org/10.1016/S1462-0758(00)00026-1.

Fiel. 2020. "Technical Catalog of Cartridge Filters and Purifiers." https://www.fielhigienizadora. com.br/wp-content/uploads/2017/12/Filtro_central_Cartucho.pdf.

Fortlev. 2019a. "Technical Catalog of Fiberglass Tank Manufacturer." https://www.fortlev. com.br/downloads/.

Fortlev. 2019b. "Technical Catalog of Polyethylene Tank Manufacturer." https://www.fortlev. com.br/downloads/.

Frischknecht, R., N. Jungbluth, H.J. Althaus, G. Doka, R. Dones, R. Hischier, S. Hellweg, et al. 2007. "Implementation of Life Cycle Impact Assessment Methods." Data v2.0, Ecoinvent report No. 3, Dübendorf, Switzerland. https://inis.iaea.org/collection/ NCLCollectionStore/_Public/41/028/41028089.pdf?r=1&r=1.

Genz, F, Tanajura, CAS, Araújo, HA. 2011. "Impacto Das Mudanças Climáticas Nas Vazões Dos Rios Pojuca, Paraguaçu e Grande - Cenários de 2070 a 2100 [Climate Change Impacts on the FLows of the Pojuca, Paraguaçu and Grande Rivers - Scenarios from 2070 to 2100]." Bahia Análise & Dados 807–824.

Ghimire, Santosh R., John M. Johnston, Wesley W. Ingwersen, and Troy R. Hawkins. 2014. "Life Cycle Assessment of Domestic and Agricultural Rainwater Harvesting Systems." Environmental Science & Technology 48 (7): 4069–4077. https://doi.org/10.1021/ es500189f.

Ghimire, Santosh R., John M. Johnston, Wesley W. Ingwersen, and Sarah Sojka. 2017. "Life Cycle Assessment of a Commercial Rainwater Harvesting System Compared with a Municipal Water Supply System." Journal of Cleaner Production 151 (May): 74–86. https://doi.org/10.1016/j.jclepro.2017.02.025.

Ghisi, Enedir. 2006. "Potential for Potable Water Savings by Using Rainwater in the Residential Sector of Brazil." Building and Environment 41 (11): 1544–1550. https:// doi.org/10.1016/j.buildenv.2005.03.018.

Ghisi, Enedir, Diego Lapolli Bressan, and Maurício Martini. 2007. "Rainwater Tank Capacity and Potential for Potable Water Savings by Using Rainwater in the Residential Sector of Southeastern Brazil." Building and Environment 42 (4): 1654–1666. https://doi. org/10.1016/j.buildenv.2006.02.007.

Gires, Auguste, Agathe Giangola-Murzyn, Jean-Baptiste Abbes, Ioulia Tchiguirinskaia, Daniel Schertzer, and Shaun Lovejoy. 2015. "Impacts of Small Scale Rainfall Variability in Urban Areas: A Case Study with 1D and 1D/2D Hydrological Models in a Multifractal Framework." Urban Water Journal 12 (8): 607–617. https://doi.org/10.1080/15730 62X.2014.923917.

Graedel, TE, llenby, BR. 1998. Design for Environment. Pearson College Division.

Guanais, A. L. R. 2015. "Avaliação Energética e Das Emissões de Gases de Efeito Estufa Do Sistema Integrado de Abastecimento de Água de Feira de Santana [Energy Evaluation and Greenhouse Gas Emissions from the Integrated Water Supply System of Feira de Santana]." Feira de Santana State University.

Guanais, Ana Luiza Rezende, Eduardo Borges Cohim, and Diego Lima Medeiros. 2017. "Avaliação Energética de Um Sistema Integrado de Abastecimento de Água [Energy Evaluation of an Integrated Water Supply System]." Engenharia Sanitaria e Ambiental 22 (6): 1187–1196. https://doi.org/10.1590/s1413-41522017146180.

Hamiche, Ait Mimoune, Amine Boudghene Stambouli, and Samir Flazi. 2016. "A Review of the Water-Energy Nexus." Renewable and Sustainable Energy Reviews 65 (November): 319–331. https://doi.org/10.1016/j.rser.2016.07.020.

Hammond, G., Jones, C. 2008. "Inventory of Carbon & Energy (ICE)." Sustainable Energy Research Team (SERT) Vision Carbon Buildings Program. https://perigordvacance. typepad.com/files/inventoryofcarbonandenergy.pdf.

Heyworth, JS, G Glonek, EJ Maynard, PA Baghurst, and J Finlay-Jones. 2006. "Consumption of Untreated Tank Rainwater and Gastroenteritis among Young Children in South Australia." International Journal of Epidemiology 35 (4): 1051–1058. https://doi. org/10.1093/ije/dyl105.

Hora, Jálvaro da, Eduardo Borges Cohim, Samuel Sipert, and Adriano Leão. 2017. "Quantitative Microbial Risk Assessment (QMRA) of Campylobacter for Roof-Harvested Rainwater Domestic Use." In Proceedings, 2:185. https://doi.org/10.3390/ecws-2-04954.

IBGE (Brazilian Institute for Geography and Statistics). 2010. "Demographic Census." Population tables for Brazil and Federation Units, Rio de Janeiro. https://www.ibge.gov. br/estatisticas/sociais/populacao/9662-censo-demografico-2010.html?=&t=o-que-e.

INMET (National Institute for Meteorology). 2018. "Rainfall Data for Feira de Santana." Historical weather data Ministry of Agriculture and Livestock. https://portal.inmet.gov. br/.

ISO (International Organization for Standardization). 2006a. "ISO 14040: Environmental Management - Life Cycle Assessment - Principles and Framework." Second edition, p. 20. https://www.iso.org/standard/37456.html.

ISO (International Organization for Standardization). 2006b. "ISO 14044: Environmental Management - Life Cycle Assessment - Requirements and Guidelines." First edition., p. 46 https://www.iso.org/standard/38498.html.

Jesus, Thiago Barbosa De, Alice Costa Kiperstok, and Eduardo Borges Cohim. 2019. "Life Cycle Assessment of Rainwater Harvesting Systems for Brazilian Semi-Arid Households." *Water and Environment Journal* 34: 1–9. https://doi.org/10.1111/wej.12464.

Leão, Adriano Souza, Samuel Alex Sipert, Diego Lima Medeiros, and Eduardo Borges Cohim. 2022. "Water Footprint of Drinking Water: The Consumptive and Degradative Use." *Journal of Cleaner Production* 355 (June): 131731. https://doi.org/10.1016/j. jclepro.2022.131731.

Marinoski, A K, and E Ghisi. 2011a. "Assessment of the Environmental Impact and Investment Feasibility Analysis of Rainwater Use in Houses." *WIT Transactions on Ecology and the Environment* 148: 391–402. https://doi.org/10.2495/RAV110361.

McDonald, Robert I., Katherine Weber, Julie Padowski, Martina Flörke, Christof Schneider, Pamela A. Green, Thomas Gleeson, et al. 2014. "Water on an Urban Planet: Urbanization and the Reach of Urban Water Infrastructure." *Global Environmental Change* 27 (July): 96–105. https://doi.org/10.1016/j.gloenvcha.2014.04.022.

McGuirk, C. Michael, Morgan D. Bazilian, and Daniel M. Kammen. 2019. "Mining Plastic: Harvesting Stored Energy in a Re-Use Revolution." *One Earth* 1 (4): 392–394. https:// doi.org/10.1016/j.oneear.2019.10.013.

Mopa. 2019. "Technical Catalog of Galvanized Steel Gutter Manufacturer." http://www.mopa. com.br/static/catalogos_downloads/75cba95c469daf40d493eb6d0d6fc17309092304. pdf.

Moreno Ruiz, E., L. Valsasina, D. FitzGerald, F. Brunner, C. Vadenbo, C. Bauer, G. Bourgault, A. Symeonidis, and G. Wernet. 2016. Documentation of changes implemented in ecoinvent database v3. 3." Ecoinvent: Zürich, Switzerland. https://forum.ecoinvent.org/files/ change_report_v3_5_20180823.pdf.

Oliveira, AKN, Ferraro Júnior, LA, Jesus, TB. 2021. "Bacia Hidrográfica Do Paraguaçu Sob Diferentes Perspectivas: Produto Técnico Educacional [Paraguaçu Watershed under Different Perspectives]." Feira de Santana. https://educapes.capes.gov. br/bitstream/capes/600537/2/Atlas_Bacia%20Hidrogr%c3%a1fica%20do%20 Paragua%c3%a7u%20sob%20diferentes%20perspectivas_P2_Alane%20Kelly%20 Nunes%20de%20OliveiraFINAL.pdf.

Oregi, Xabat, Patxi Hernandez, Cristina Gazulla, and Marina Isasa. 2015. "Integrating Simplified and Full Life Cycle Approaches in Decision Making for Building Energy Refurbishment: Benefits and Barriers." *Buildings* 5 (2): 354–380. https://doi.org/10.3390/buildings5020354.

Parkes, C., Kershaw, H., Hart, J., Sibille, R., & Grant, Z. 2010. Energy and carbon implications of rainwater harvesting and greywater recycling. Final Report, Science Project Number: SC090018, Environment Agency, Bristol. https://assets.publishing.service.gov. uk/media/5a7bfda6e5274a7318b905b9/scho0610bsmq-e-e.pdf.

Peel, M. C., B. L. Finlayson, and T. A. McMahon. 2007. "Updated World Map of the Köppen-Geiger Climate Classification." *Hydrology and Earth System Sciences* 11 (5): 1633–44. https://doi.org/10.5194/hess-11-1633-2007.

Pentair. 2014. "Technical Catalog of Cartridge Filters and Purifiers." https://www.inovahouse.com.br/wp-content/uploads/2018/04/cat-2014-baixa-res.pdf.

Richter, Brian D., Mary Elizabeth Blount, Cara Bottorff, Holly E. Brooks, Amanda Demmerle, Brittany L. Gardner, Haley Herrmann, et al. 2018. "Assessing the Sustainability of Urban Water Supply Systems." *Journal of American Water Works Association* 110 (2): 40–47. https://doi.org/10.1002/awwa.1002.

Rodrigo, S., Ledar, K., Sinclair, M. 2009. "Quality of Stored Rainwater Used for Drinking in Metropolitan South Australia: Research Report 84." https://research.monash.edu/en/publications/quality-of-stored-rainwater-used-for-drinking-in-metropolitan-sou.

Sánchez, A.S., E. Cohim, and R.A. Kalid. 2015. "A Review on Physicochemical and Microbiological Contamination of Roof-Harvested Rainwater in Urban Areas." *Sustainability of Water Quality and Ecology* 6 (September): 119–137. https://doi.org/10.1016/j.swaqe.2015.04.002.

Schoen, Mary E., and Jay Garland. 2017. "Review of Pathogen Treatment Reductions for Onsite Non-Potable Reuse of Alternative Source Waters." *Microbial Risk Analysis* 5 (April): 25–31. https://doi.org/10.1016/j.mran.2015.10.001.

SEEG (Greenhouse Gas Emissions Estimation System for Brazil). 2020. "State of Bahia: Emissions Profile of 2017." 2020. https://plataforma.seeg.eco.br/territories/bahia/card?year=2017&cities=false.

Shockmetais. 2019. "Technical Catalog of Aluminum Gutter Manufacturer." https://shockmetais.com.br/produtos/aluminio.

da Silva, Mariana P., Jorge González, Bruno B. F. da Costa, Claudia Garrido, Carlos A. P. Soares, and Assed N. Haddad. 2023. "Environmental Impacts of Rainwater Harvesting Systems in Urban Areas Applying Life Cycle Assessment-LCA." *Eng* 4 (2): 1127–1143. https://doi.org/10.3390/eng4020065.

Silva, Michelle, Josefa Eliane Santana de Siqueira Pinto, and Francisco Jablinski Castelhano. 2022. "Análise Da Variabilidade Pluvial e Sua Contribuição Para o Estudo Do Clima Urbano Do Município de Feira de Santana-BA." *Geopauta* 6 (April): e10251. https://doi.org/10.22481/rg.v6.e2022.e10251.

SIM (Municipal Information System). 2017. "Uso de Eletricidade Por Setor de Consumo [Electricity Use by User Sector]." 2017. https://sim.sei.ba.gov.br/sim/index.wsp.

SINAPI CEF (National System for Research of Costs and Indices of Civil Construction from Caixa Econômica Federal). 2019. "Datasets of Economic Indicators for Construction Materials, Machinery and Labor per Service."

Springway. 2020. "Technical Catalog of Cartridge Filters and Purifiers." https://springway.com.br/wp-content/uploads/2020/08/catalogospringway.pdf.

Stec, Agnieszka, and Martina Zeleňáková. 2019. "An Analysis of the Effectiveness of Two Rainwater Harvesting Systems Located in Central Eastern Europe." *Water* 11 (3): 458. https://doi.org/10.3390/w11030458.

Tanajura, Clemente Augusto Souza, Fernando Genz, and Heráclio Alves de Araújo. 2010. "Mudanças Climáticas e Recursos Hídricos Na Bahia: Validação Da Simulação Do Clima Presente Do HadRM3P e Comparação Com Os Cenários A2 e B2 Para 2070-2100." *Revista Brasileira de Meteorologia* 25 (3): 345–358. https://doi.org/10.1590/S0102-77862010000300006.

Tigre. 2019. "Technical Catalog of PVC Pipes Manufacturer." https://www.tigre.com.br/catalogos-tecnicos.

Valdez, M. Carmen, Ilan Adler, Mark Barrett, Ricardo Ochoa, and Angel Pérez. 2016. "The Water-Energy-Carbon Nexus: Optimising Rainwater Harvesting in Mexico City." *Environmental Processes* 3 (2): 307–323. https://doi.org/10.1007/s40710-016-0138-2.

Valério, ELS, Fragoso, CR. 2015. "Avaliação Dos Efeitos de Mudanças Climáticas No Regime Hidrológico Da Bacia Do Rio Paraguaçu, BA. [Evaluation of the Effects of Climate Change on the Hydrological Response of the Paraguaçu River Basin]." *Rev. Bras. Recur. Hídricos* 20, 872–887. https://abrh.s3.sa-east-1.amazonaws.com/Sumarios/156/1e7c118 e57ba51b7bce58356f398da1d_ba28e48c208d8cdb9817500bf2e89fc1.pdf.

Vieira, A. S., and E. Ghisi. 2016. "Water-Energy Nexus in Houses in Brazil: Comparing Rainwater and Gray Water Use with a Centralized System." *Water Supply* 16 (2): 274–283. https://doi.org/10.2166/ws.2015.137.

Voinov, Alexey, and Hal Cardwell. 2009. "The Energy-Water Nexus: Why Should We Care?" *Journal of Contemporary Water Research & Education* 143 (1): 17–29. https://doi. org/10.1111/j.1936-704X.2009.00061.x.

Ward, Sarah, David Butler, and Fayyaz Ali Memon. 2012. "Benchmarking Energy Consumption and CO_2 Emissions from Rainwater-harvesting Systems: An Improved Method by Proxy." *Water and Environment Journal* 26 (2): 184–190. https://doi. org/10.1111/j.1747-6593.2011.00279.x.

Wernet, Gregor, Christian Bauer, Bernhard Steubing, Jürgen Reinhard, Emilia Moreno-Ruiz, and Bo Weidema. 2016. "The Ecoinvent Database Version 3 (Part I): Overview and Methodology." *The International Journal of Life Cycle Assessment* 21 (9): 1218–1230. https://doi.org/10.1007/s11367-016-1087-8.

WULCA. 2016. "Download AWARE Factors." https://Wulca-Waterlca.Org/Aware/Download-Aware-Factors/. 2016.

3 Optimisation of Rainwater Harvesting System Design through Remote Sensing and Modelling

Ilan Adler, Jordan Silverman, and David Ruvubi

3.1 INTRODUCTION

Accessibility and affordability of clean water are regarded as human rights, which are underpinned by the UN Sustainable Development Goal (SDG 6), yet management of water resources has been challenging due to diverse factors, including industrialisation, increasing population and human activities. The water crisis has been ranked among the top five global risks in terms of likelihood and impact on the global population, and over 2 billion people experience high water stress, mostly in developing countries (UN World Water Report, 2019). To address the problem of clean water access, a decentralised, treated rainwater harvesting (RWH) system could be a viable alternative water supply system.

Rainwater harvesting is a traditional and sustainable water collection technique used for potable and non-potable purposes both in residential and commercial buildings (Rahman et al., 2014). It is also associated with the additional benefits of mitigating flood risk and soil erosion (Gwenzi & Nyamdazawo, 2014), helping adapt to climate change and minimising its impact (Han & Mun, 2011; Fernandes et al., 2015). Accordingly, an RWH system plays a pivotal role in the circular economy and restoration of the water cycle, as the cost caused by flooding and the demand for water can be reduced, aquifers can be recharged and wastewater treatment can be improved (Espindola et al., 2018).

However, conventional RWH systems possess drawbacks. Many RWH systems fail to operate in the long run due to a lack of regular maintenance, training and replacement parts, which can affect access and the quality of water, leading to risks to public health (Adler et al., 2014). In addition, available information regarding the design of rainwater harvesting systems is largely based on empirical data such as historical average annual precipitation (Krishna, 2005), which may not reflect actual trends or variations due to climate change, local rainfall patterns and microclimates, which may vary from one site to another in the same area.

DOI: 10.1201/9781032638102-4

To overcome the limitations of RWH systems, several studies have applied Real-time Control Systems (RCS), which allow remote control of systems via a wireless network. However, incorporating the RCS with an RWH system is relatively less explored, with limited exceptions such as Seoul, South Korea (Han & Mun, 2011), North Carolina, United States (Gee & Hunt, 2016) and Melbourne, Australia (Xu et al., 2018).

This chapter describes the application of remote monitoring techniques in RWH systems. Using remote sensing and monitoring in real-time with low-cost equipment, the study addresses the drawbacks of conventional versus RCS incorporated RWH systems regarding the estimation of rainwater collection, affordability, maintenance and design.

The scope of the study includes constructing a real-time monitoring network for an RWH system to evaluate correlation factors, optimise rain collection efficiency and optimise hardware design as well as the predictive model of rainwater collection. The research objectives were to:

a. Implement fully functional real-time monitoring of an RWH system that can be replicated in low- and lower-middle-income countries using low-cost equipment.
b. Provide monitoring and calibration for the field trial, with recommendations on design optimisation for future work.
c. Calibrate predictive models of rainwater harvesting by comparing existing models with real-time experimental data and determining a precise runoff coefficient.
d. Develop operation and maintenance protocols to deal with the failures and operations of typical RWH systems.

3.2   LITERATURE REVIEW

3.2.1   COMPONENTS OF AN RWH SYSTEM

In general, there are seven main components in an RWH system. Specific components can vary according to different functions, scales and types of systems. According to Mishra (2019) and Susilo (2018), the main components are listed in Figure 3.1, which shows the main structure of an RWH system.

3.2.2   DETERMINING RUNOFF COEFFICIENT

The runoff coefficient (RC) is a dimensionless value that approximates the portion of rainfall that becomes runoff, bearing in mind losses due to spillage, leakage, catchment surface wetting and evaporation (Farreny et al., 2011). It is a ratio calculated between the volume of rainwater that runs off a surface and the volume of rainwater that falls on that surface. RC is important to predict the potential runoff surface water that can be conveyed to a rainwater storage system (Miller, 2017). Biswas and Mandal (2014) denote the importance of efficiency in collecting runoff in RWH systems to the amount of water captured in the tank. For instance, a runoff coefficient of 0.7 means that 70% of the rainfall that fell on the rooftop surface is collected in the

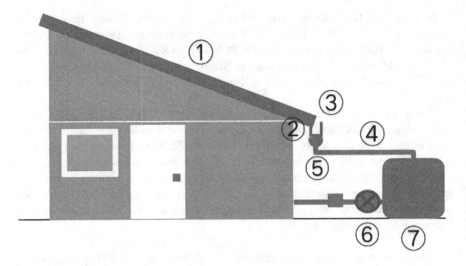

FIGURE 3.1 Components of an RWH system (Susilo, 2018). (1) Rooftop: A surface such as a rooftop or a paved flooring surface where rainwater could be collected. (2) Coarse mesh: to prevent the passage of debris or solids going into pipes. (3) Gutters: Connected to the roof to deliver roof water into the first flush. (4) Conduit: Pipes or drains to carry water from the gutter into the system. (5) First flush: to prevent pollutants on the roof from going into the tank. (6) Filter: to remove suspended pollutants in roof water. (7) Storage tank: to store rainwater collected.

tank. Accordingly, a higher RC refers to a higher volume of rainwater collected in the tank, which means a higher efficiency of rainwater collection.

Farreny et al. (2011) analysed four roof types: three sloping roofs of clay tiles, metal sheets and polycarbonate plastic and one flat gravel roof. They observed that sloping smooth roofs (RC > 0.90) harvest up to circa 50% more rainwater than flat rough roofs (RC = 0.62). Criteria selection for RC is recommended to estimate rainwater collection efficiently.

The review of other RC estimates that have been studied by various researchers is indicated in Table 3.1. In the UNICEF manual for construction and maintenance of household-based rooftop water harvesting systems, Kumar (2004) estimated the RC for concrete to be 0.7. However, Lancaster (2006b) estimates that the RC is 0.9 for concrete and asphalt roof types with a slope. This variance should be considered in determining RC given that the evaporation rate, wind, slope and roughness of the roof in RWH are essential in urban/rural planning to promote rainwater as an alternative supply while mitigating flooding and water scarcity (Kumar 2004; Lancaster 2006a; Ward et al., 2010).

3.2.3 RAINWATER COLLECTION POTENTIAL

Various formulae are available to estimate the amount of rainwater collected in the storage tank from the rooftop (Lancaster, 2006b; Kim et al., 2019; Farreny et al., 2011). Two formulae that are most commonly used in estimating the rainwater

TABLE 3.1
Runoff Coefficient Estimates of Different Roof Types

Types of Roof	RC	Reference
	General	
Galvanised Iron Sheet	0.9	Kumar (2004)
Tiled roof	0.75	Kumar (2004)
Concrete	0.7	Kumar (2004)
Asbestos Sheet	0.5–0.6	Kumar (2004)
	Sloping roofs	
Concrete/asphalt	0.9	Lancaster (2006b)
Metal	0.95	Lancaster (2006b)
Metal	0.81–0.84	Liaw and Tsai (2004)
Aluminium	0.7	Ward et al (2010)
	Flat roofs	
Bituminous	0.7	Ward et al (2010)
Gravel	0.8–0.85	Lancaster (2006b)
Level-cement	0.81	Liaw and Tsai (2004)

TABLE 3.2
Different Estimations for Calculating Collected Water (Lancaster 2006b)

Estimation	Formulas	Variables
Theoretical (Rooftop Collection based on rainfall)	$Q = A \times RC \times R$	Q: Water collected (L) A: Catchment Area (m²) RC: Catchment runoff coefficient (dimensionless) R: Rainfall data from weather station (mm)
Experimental (Water volume collected in tank)	$Q = \triangle H \times TA$	Q: Water collected (L) $\triangle$H: Change in height of water column (mm) TA: Tank surface area (m²)

collection potential are selected for the detailed review: theoretical and experimental estimation. The estimation approach, equation and variables used for each estimation are summarised in Table 3.2. The theoretical estimation is dependent on the RC, rainfall data and catchment area, whereas the experimental estimation is based on an experiment that measures the water level inside the tank to calculate the rainwater collection.

3.2.4 Remote Monitoring and Data Transmission

Remote monitoring systems, supporting the observation of activity, can be realised by using Internet of Things (IoT) technology, which provides the interconnection of components with the capabilities of built-in computing, communication and sensing

(Atzori et al., 2010). Once remote monitoring is incorporated with an RWH system, the accuracy and precision of the runoff coefficient can be improved; accordingly, it contributes to developing the optimal design of RWH systems (Adham et al., 2016). The cost, time and labour required to observe the status of the RWH system on-site can be saved by supervising the system online in remote areas.

To coordinate the tasks between systems and process data, a microcontroller unit (MCU) – a stripped-down, miniature computer designed to control the operating system – can be utilised (Crysdian, 2017). In general, an MCU includes a processor, memory and input/output (I/O) peripherals on a single chip, and it can be programmed to command specific tasks to the connected device via a programming platform. The data can be saved to the onboard memory of the MCU, but storage devices (e.g., SD-cards) can be added if extra storage capacity is required.

After the data is collected, a data transmission mechanism is required to allow monitoring the data remotely. To identify the most suitable data transmission technology for our RWH system, the characteristics of several options, including 2G/3G, Bluetooth, LoRa, SigFox and Wi-Fi, were investigated. LoRa and SigFox are two optimal technologies for the RWH system, as these can send the data over a long distance with medium cost and low power consumption. The low power consumption means that these devices can run for up to eight years on a single AA battery when optimised correctly. Furthermore, both technologies were adopted for their operability on licence-free frequencies, meaning they can freely operate wireless networks without requesting permission from an issuing authority, provided the operation guidelines are adhered to.

One of the drawbacks of using LoRa or Sigfox technology is that when the data reaches from the remote monitoring device to a centralised gateway, that gateway still requires access to the internet via a wired network or using mobile technology. When deploying remote monitoring systems, these technologies become financially viable and practical when there are several systems being monitored within a few kilometres radius and there is access to an internet connection within that radius.

3.3 METHODOLOGY

An RWH system with remote monitoring using low-cost equipment was developed to test the functionality of the system in real-time. The primary study objective was to determine a more precise RC compared to the RC = 0.9 that the other literature found to be most suitable based on the material of the rooftop (asphalt) and slope (28.36°). This study focused on the optimisation of the system design and data monitoring to allow replicating the system in various climatic conditions and regions, especially in low- and lower-middle-income countries. The design of the storage tank was modified by adding an overflow outlet, a transparent tube and a metal plate. The remote monitoring network was optimised by incorporating an online platform (TeamViewer) for data access and developing firmware running on the MCU for efficient sensor sampling and data transmission. An ultrasonic sensor was used to detect the water level in the tank. This type of sensor was a relatively low-cost device that could measure water depth with great accuracy. It is worth noting that pressure transducers were considered a viable methodology for calculating the amount of water in the tank; however, they would be hard to operationalise in underground cisterns, which are common in RWH systems. After calibration of the ultrasonic sensor, the volume of rainwater collected was calculated based on theoretical and experimental estimations (Section 2.3). Rainfall data from a weather station was used to calculate the precise RC, and other weather variables were deployed to identify the correlation between each factor and the rainwater collection efficiency.

3.3.1 STUDY AREA CHARACTERISTICS

To base the modelling and remote monitoring study on reality, the experiment was conducted in London in 2019–20 at the 'Eco-hub', which was located at the Pedestrian Accessibility Movement Environmental Laboratory (PAMELA) in Tufnell Park, London.[1] The surrounding weather stations near the Eco-hub were investigated, and the three closest stations were the Heathrow weather station, Oxford weather station and Cambridge NIAB weather station. 15, 50 and 45 miles away from the Eco-hub, respectively (Historic Station Data, 2020).

As precipitation is the key factor in rainwater collection efficiency in RWH systems, it was important to investigate the historical rainfall data to determine the suitability and potential of the system. Given the fact that rainfall events vary year to year and season to season, the investigation looked at a substantial period. The historical data of UK rainfall from 1981 to 2010 (30 years) was examined as shown in Figure 3.2. The rainfall from 1981 to 2010 fluctuated between 820 and 1,480 mm per year, with an average rainfall of 1,150 mm. The above data highlighted that there was sufficient rainfall in London. Notwithstanding London's prolonged rainy season, the impact of the dry season and the fluctuation in annual rainfall on the RWH

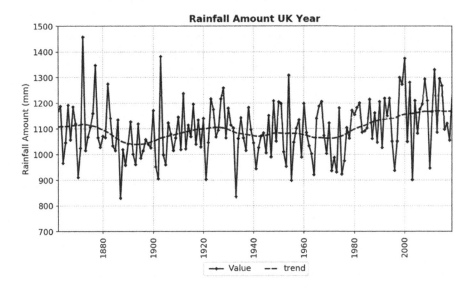

FIGURE 3.2 Historical data on the amount of the rainfall in the UK (Met Office, 2020a, b, c).

system still needs to be controlled. Therefore, it was decided that the RWH system could be effectively tested and implemented to harvest the rainwater in the selected catchment area.

3.3.2 OVERVIEW OF THE RWH SYSTEM

Figure 3.3(a) and (b) represent the dimensions and details of the components from the top-view and the side-view of the RWH system. As shown in Figure 3.3(a), the main components of the RWH system, such as a rooftop, gutter and tank inside the shed, were located 10 m from the office building, functioning as the control centre of the system, and the weather station was installed 2 m away from the shed. Figure 3.3(a) shows the details of the system's components from the side view and the interaction

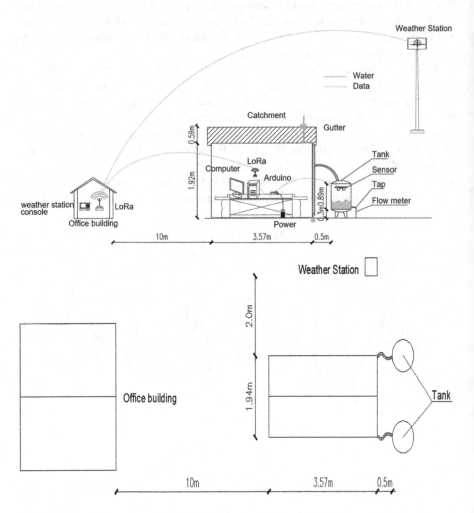

FIGURE 3.3 (a) Top view of the RWH system at the Eco-hub. (b) Side view. *The distance between the RWH system and the office building (10 m) is not drawn to scale.

between the system, data and rainwater. The pathways for data transmission and for rainwater are illustrated with light and thick lines, respectively. The rainwater was captured from the catchment area and sent into the storage tank through gutters and pipes. An ultrasonic sensor fixed on the top of the storage tank enabled monitoring of the water volume by detecting the water level collected in the tank. The sensor was read by an Arduino, and the firmware calculated the volume of water in the tank. The data was then transmitted to a laptop in the office building via LoRa.

3.3.3 Physical Properties of the RWH System

As illustrated in Table 3.3, each component of the RWH system was measured to provide the physical parameters needed for design and further analysis. The experiment was performed with a single tank, with only half of the two-sided pitched roof used for rainwater collection (the shed was symmetrical). The catchment area was thus calculated to be the footprint of a single side of the roof, as shown below.

3.3.4 Hardware Improvements

Setting up a functional and integrated prototype of the RWH system required the following improvements to achieve better accuracy for measuring water level and volume in the water tank. A transparent tube was affixed to the side of the opaque

TABLE 3.3
Physical Properties of the RWH System

Component	Dimensions
Left half of the rooftop *See Figure 3.3*	Height (H): 0.59 m Width (W): 0.97 m Length (L): 3.58 m Catchment area (A) $= W \times L = 3.47\,m^2$ Roof angle (θ): 28.4° Material: Asphalt
Storage Tank *See Figures 3.4 and 3.8*	Tank diameter (D) :0.40 m Cross-sectional area: 0.12 m^2 Height of stand (base): 0.36 m Total Height including base: 1.24 m Height of transparent tube used for measuring water column: 0.6 m Capacity: 100 L (Material: plastic)
Gutter	Diameter (D): 0.12 m Cross-section shape: semicircle Material: Polyvinyl Chloride (PVC)
Downspouts	Height: 1.82 m Diameter: 0.068 m Material: PVC

storage tank to observe the amount of rainwater collected in the tank (Figure 3.4). The tube was calibrated to determine the water volume in the rectangular section of the tank, disregarding the top and bottom sections. Moreover, the overflow outlet was fitted to prevent the ultrasonic water level sensor from being flooded during the heavy rain event.

As shown in Figure 3.5, the wooden plate that had been previously fitted to hold the ultrasonic sensor on the top of the tank was replaced with a secure metal plate to mitigate the risk of the sensors being damaged by water, as a wood plate could potentially absorb the moisture and dampen the sensor. Moreover, the customised metal plate was also securely fitted with screws to avoid any movements whilst measuring the water level.

Before   After

FIGURE 3.4   Design improvements of the tank by adding a transparent tube and an overflow outlet

Before   After

FIGURE 3.5   Improvements made to the water level sensor holder.

3.3.5 Configuration of the Remote Monitoring System

The components of the remote monitoring system were selected based on three important factors: cost, power consumption and accessibility to users. As illustrated in Figure 3.6, the remote monitoring process of the RWH system consisted of five main stages to provide real-time monitoring measurements to the user: sensor input, system control, data transmission, analysis of the data and the user interface.

3.3.5.1 Sensor Input

To detect the current status of the rainwater collection in the storage tank, an ultrasonic sensor (HC-SR504) was deployed to measure the distance from the sensor installed on the top of the tank to the water level accumulated in the tank. An ultrasonic sensor detected the water level by sending out the sound waves when commanded to by the microcontroller via the trigger pin, followed by returning the results to the microcontroller by setting the echo pin to high (5 V) for a length of time proportional to the time between sending and receiving the sound waves, therefore determining the distance between the sensor and the water level (Prima et al., 2017). The specifications of the ultrasonic sensor HC-SR504 are summarised in Table 3.4.

3.3.5.2 System Control

For the remote monitoring system control, an Arduino, which is a board mainly made of an MCU, a power port, analogue/digital pins and a USB connector, was used to command the connected devices (e.g., ultrasonic sensors, SD cards and LoRa modules) to perform actions. The command was executed by writing firmware using the Arduino Integrated Development Environment (IDE). The firmware was compiled and uploaded to the MCU via USB. It ran continuously, executing the functions in a loop

FIGURE 3.6 A flow chart of the RWH remote monitoring systems.

TABLE 3.4

The Specification of HC-SR504 Sensor (Kruger, 2017)

Parameter	Specification
Working voltage/current	DC 5 V/15 mA
Working frequency	40 Hz
Max–min range	2–400 cm
Measuring angle	15°
Dimension	45 × 20 × 15 mm

whilst it stayed connected to the power source. The Arduino was selected because it provided a simple programming environment, open source, cross-platform (e.g., Windows, Macintosh OSX and Linux) with low-cost (Dada et al., 2018; Arduino, 2019).

Due to the limited storage capacity of the Arduino board, an SD card with a capacity of 8 GB was inserted to extend the backup storage capacity using a micro-SD card module for Arduino. After inserting the SD card into the SD card module for Arduino, the SD card module was connected to the Arduino Uno board using a serial peripheral interface (SPI).

To supply the power for the system control, a 5 V battery pack (Anker Astro E1) was used. The battery is compact and portable in design, and it could be fully recharged in 5 hours by USB cable. Once the battery was fully charged, it ran for about 2–3 days without recharging, in a scenario where the data was being transmitted every 5 minutes using LoRa and stored on the SD card. The Arduino was a prototyping board and had not been optimised for super low power consumption, as would be the case in a production version.

3.3.5.3   Data Transmission

For this experiment, LoRa was used to transmit data wirelessly to a nearby gateway because it is extensively used in many real-time monitoring systems, such as monitoring and tracking maritime activities (Li et al., 2017), meteorological information display systems (Reda et al., 2018), smart irrigation systems and solid waste management systems (Bharadwaj et al., 2016). Moreover, the suitability and functionality of LoRa with an RWH system had been previously tested at University College London (UCL). LoRa consists of two modules with antennas: a transmitting module and a receiving module. The transmitting module with antenna was interfaced into the Arduino Uno board, and the receiving module with antenna was connected to the gateway device (in this case, a laptop) to establish point-to-point communication. To send the data from the Arduino Uno to the monitoring device, the Arduino was programmed to communicate with a LoRa module over a serial interface, feeding data on a regular basis. The LoRa module operated the wireless protocol and modulation on its own processor. Figure 3.7 shows a diagram of how the components were connected to form the remote monitoring system.

3.3.5.4   Remote Monitoring and Troubleshooting

The data transferred via LoRa was monitored on the laptop in the office building. The receiving module and antenna were connected to the laptop and the measurement data could be monitored using the Serial Port Utility software downloaded on the laptop. The limitation to the remote monitoring was that LoRa could only transfer data within 1–3 miles. To overcome this limitation and enable remote monitoring of real-time data from diverse locations, a software application called TeamViewer was utilised. TeamViewer is a comprehensive software application used for both industry and the private sector that allows accessing devices (e.g., PCs, mobiles and servers) located anywhere in the world. Remote access was achieved by installing a TeamViewer application on both devices. The data was secure as the connection and access could only be made with the given IP and password, which change every time the device is turned off.

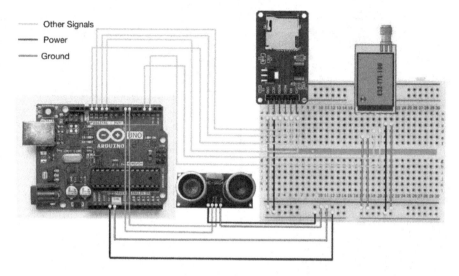

FIGURE 3.7 The circuit diagram of an Arduino Uno, ultrasonic sensor, SD card and LoRa module using breadboard.

To achieve the objectives of the project, not only the real-time data from the tank was collected, but also the weather information was gathered using a weather station. WeatherLink was used to measure precipitation, wind speed, solar radiation and temperature. The weather station (Davis Vantage Pro 2) had already been installed adjacent to the shed to provide accurate weather data at the point of installation. The weather data was continuously monitored via both the console and the laptop with installed WeatherLink software.

3.3.6 ESTIMATION OF THE RAINWATER COLLECTION VOLUME

The attached transparent tube on the water tank had to be calibrated for the purpose of determining the water volume collected inside the tank. Then 5 L of water was poured into the tank, and the corresponding level in centimetres on the tube was measured. For this experiment, 5 L of water equated to 4 cm in the storage tank. Thus, the transparent tube was marked with a 4 cm increment for each segment. For the experiment, the volume of the point-of-interest was determined to be the rectangular area as shown in Figure 3.8. The top and bottom parts of the tank with the curvature were disregarded to allow an accurate estimation of the volume. Therefore, the volume was calculated from the baseline (16.5 cm) to the maximum water level (78.5 cm), with a middle part of the tank measuring 62 cm as the height for the calculation. The total volume capacity of the rectangular section was estimated at 77.5 L because 4 cm was equivalent to 5 L of the volume inside the tank. To make this calculation possible, the water level at the baseline was always maintained whenever the tank was drained. Based on these assumptions, the volume was calculated by the following (Equation 3.1):

$$\text{Volume} = (13\,\text{cm} + 62\,\text{cm} - \text{distance to water level (cm)}) * 1{,}270\,\text{cm}^2 \quad (3.1)$$

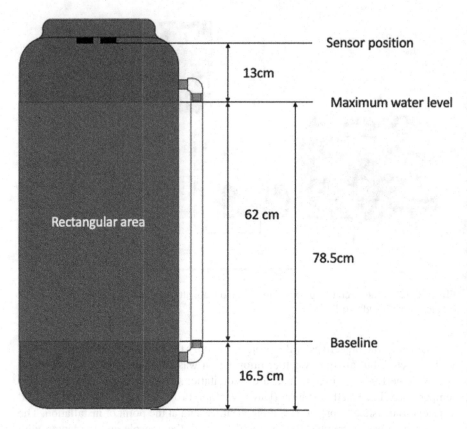

FIGURE 3.8 Dimension and representation of the point-of-interest area used for tank volume calculations.

3.3.7 CALCULATION OF RAINWATER COLLECTION POTENTIAL

For a more representative and meaningful theoretical estimation of the rainwater collection, the theoretical estimation formula was manipulated to calculate the average volume per 5 minutes. For the theoretical estimation, the RC for the system was assumed to be 0.9 based on the material and the slope of the roof (Table 3.1). The amount of water collected per 5 minutes (QM) was calculated as follows (Equation 3.2):

$$Q_M = \frac{\sum_{M=1}^{5} A \times RC \times R_M}{5} \tag{3.2}$$

where:

QM   Average volume collected in every 5 minutes (L)
A   Catchment Area (m²)
RC   Catchment runoff coefficient (dimensionless)
RM   Rainfall in 5 minutes (mm)

For the tank collection estimation, the volume of rainwater was calculated using Equation (3.3). The water level was measured every 5 minutes, which equated to the time interval set for the rooftop rainfall estimation.

$$Q = oH \times TA \qquad (3.3)$$

where:

Q   Water collected (L)

oH   Change in change in height of water level (mm)

TA   Tank surface area (m²)

3.3.8 VALIDATION OF THE ULTRASONIC SENSOR

The accuracy of the ultrasonic sensor was validated by comparing the values obtained via Arduino to the manual readings using the transparent tube. The test was repeated ten times with different water levels in the rectangular area of the tank, from a minimum height of 13 cm to a maximum of 75 cm. The average error of the sensor is observed to be 1.31 cm, with a correlation to the manual readings of 0.998. Such a high correlation and low average error ensure the reliance of the water level measurement for further analysis.

3.4 RESULTS AND DISCUSSION

The RWH system, with remote monitoring technology using a low-cost water level sensor, gave us empirical data on the rainwater collected in the storage tank, allowing a comparison to the theoretical estimation. The experimental data on the volume of water in the tank was more accurate, even in detecting small increases during light rainfall events (<0.25 mm per 5 minutes). Thus, the RC determined using the experimental estimation can be utilised to improve the rainwater collection efficiency model and provide a more accurate prediction of future rainwater collection. The correlation of RC and environmental factors other than rainfall, such as wind direction, humidity and solar radiance, was analysed, with wind direction showing the highest correlation, but is not reported here due to a limited number of data sets (three rainy days), making it hard to conclude whether any tendencies are reliable. The correlation analysis for these other variables and environmental factors can be improved in the future by adding more data to the model and using a more controlled environment.

3.4.1 CALCULATING THE RC

Table 3.5 shows the calculation process for three rainfall events to get the precise RC for the RWH system. The amount of rainfall data captured from the local weather station seemed to be consistent at 0.25 mm. This might have been due to light rainfall events throughout the three days of the experiment and the resolution of the data that the WeatherLink equipment was able to measure. The experimental volume data is calculated by using Equation 3.3 with 5-minute intervals. As Equations 3.2

TABLE 3.5

Data Collected and RC Calculation from Three Separate Rain Events

Day	Minutes of Rainfall	Rainfall (mm)	Rainfall on Catchment Area (L)	Volume Collected in Tank (L)	RC
4 March 2020	210	2.75	9.57	8.34	0.87
5 March 2020	305	11.78	40.98	35.68	0.87
9 March 2020	80	1.25	4.34	3.81	0.88

and 3.3 calculate the volume (Q) in 5 minutes, the experimental volume difference in 5 minutes is utilised to get the RC. By setting the Q equal to the theoretical (Q=A×RC×rainfall) and experimental volumes (Q=oH×TA), the RC for each 5 minutes of interval is calculated by using the equation: RC=(oH×TA) / (A×rainfall). The area of the roof was 3.47 m² so using this equation, we were able to calculate the RC as seen in Table 3.5. The average RC measured was 0.87, which is close to the theoretical RC (Table 3.1) of 0.9 (Figure 3.9).

The graphs show the difference between the rainwater that fell on the catchment area and the rainwater collected in the tank in the three rain events. The result shows that not all the water falling on the catchment was collected into the tank. This is consistent with the calculation of the RC being on average 0.87 across all rainfall events calculated in Table 3.5.

3.4.2 MAINTENANCE OPERATION PROTOCOL

Maintenance Operation Protocol (MOP) is generally used for utility services such as uploading and downloading system software, remote testing and problem diagnosis. As this experiment included a prototype version of the remote monitoring system, there were many issues encountered during the setup process. A final product would be more robust before deploying to remote locations. However, Figure 3.10 shows a protocol to guide the operator through the necessary actions to troubleshoot the failures of the remote monitoring system, including data transmission and hardware.

The actions are divided into three steps. The first call is to use a low-cost or free open resource support and monitoring tool such as TeamViewer to access the gateway on-site in case the data transmission is lost. The Internet connection and the monitoring device can be checked and rebooted if necessary. If there is no data, a communication failure is suspected between the LoRa gateway and the microcontroller, the Arduino Uno, due to the following potential system failures: no power supply from the battery, loose connections or a sensor failure. Other necessary maintenance requirements are regular checks for debris and leaves that may cause obstruction in the rainwater conveyance from the rooftop through the gutter and pipes into the tank. On occasion when the data is remotely inaccessible, a site visit is necessary to physically examine the system and ascertain the causes of failure, as discussed above. The faulty components are either repaired or replaced. However, additional investigations may be required, depending on the complexity of the RWH system.

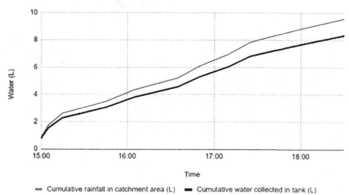

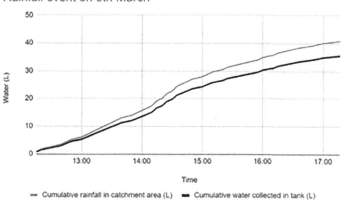

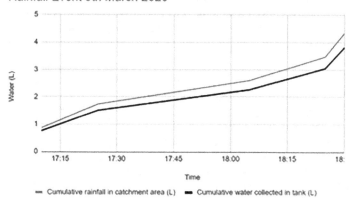

FIGURE 3.9 A plot of data from the WeatherLink station of rainfall and measured water increase in the collection tank.

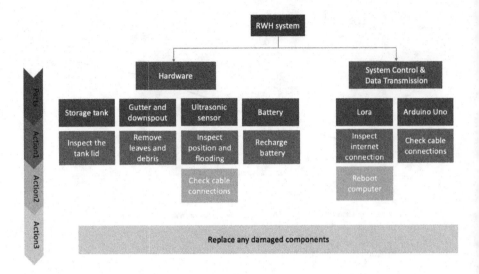

FIGURE 3.10   Hierarchical diagram of the maintenance operation protocol.

3.5   CONCLUSION AND RECOMMENDATIONS

The remote monitoring system developed in this study provides an initial prototype for practical solutions to monitor RWH systems in distant areas. The data can be used for maintenance and troubleshooting (i.e. if there is a big difference between the theoretical amount of rainwater that should have been collected vs. the observed value), as well as improving design efficiency for specific areas with real-time, site-specific data being collected and analysed vs. average historical values from weather stations that could be located far from the site or with incomplete datasets.

The experiments performed as part of this research provide a solid basis for future studies. The RC of the developed RWH system in London was determined to be 0.87, which falls in line with theoretical estimates according to roof characteristics. The data and methodology developed in the project can also be used as a baseline to continue optimising the RWH system, up to the level where it could be commercialised.

The following recommendations are proposed in order to improve the system

- Replace battery packs with solar panels for a renewable power source.
- Improve the power efficiency of the remote monitoring device.
- Reduce cost and improve the performance of the monitoring device, e.g., lower-cost units with better performance, such as the Raspberry Pi, can be applied instead of a laptop. Alternatively, data transmission devices and cloud storage can substitute for the use of an onsite processor altogether.
- Add a water quality testing and filtering system to test water quality parameters such as pH, turbidity and conductivity to ensure the quality of the water collected is suitable.
- Test different types and slopes of rooftop to ensure that the most efficient roof types and slopes are incorporated into the RWH system.

- A future step could be to add control elements (i.e. automated or remote-controlled valves), which could help improve the efficiency of the system beyond only monitoring.

ACKNOWLEDGEMENTS

The authors would like to thank Jaemee Lee, Susu Li and Yang Wang for their excellent experimental work as part of their MSc at UCL, and to Dr. Berill Takacs for her invaluable help in the editing process.

NOTE

1 It was later relocated to UCL-SAIL (Sustainable Agricultural Innovation Lab), based in Harpenden, UK.

REFERENCES

Adler, I., Campos, L., and Bell, S., 2014. Community participation in decentralised rainwater systems: a Mexican case study. In *Alternative Water Supply Systems*, A. M. Fayyaz & S. Ward, eds. Exeter: International Water Association (IWA)

Atzori, L., Iera, A. and Morabito, G., 2010. The Internet of Things: A survey. *Computer Networks*, 54(15), pp. 2787–2805.

Bharadwaj, A. S., Rego, R. and Chowdhury, A., 2016. IOT Based Application for Smart City Implementation. *International Journal of Modern Trends in Engineering & Research*, 3(9), pp. 146–149.

Biswas, B. and Mandal, B., 2014. Construction and Evaluation of Rainwater Harvesting System for Domestic Use in a Remote and Rural Area of Khulna, Bangladesh. *International Scholarly Research Notices*, 2014, pp. 1–6.

Gee, K. and Hunt, W., 2016. Enhancing Stormwater Management Benefits of Rainwater Harvesting via Innovative Technologies. *Journal of Environmental Engineering*, 142(8), p. 04016039.

Gwenzi, W. and Nyamadzawo, G., 2014. Hydrological Impacts of Urbanization and Urban Roof Water Harvesting in Water-limited Catchments: A Review. *Environmental Processes*, 1(4), pp. 573–593.

Han, M. and Mun, J., 2011. Operational Data of the Star City Rainwater Harvesting System and Its Role as a Climate Change Adaptation and a Social Influence. *Water Science and Technology*, 63(12), pp. 2796–2801.

Kumar, M., Agarwal, A. and Bali, R., 2008. Delineation of Potential Sites for Water Harvesting Structures Using Remote Sensing and GIS. *Journal of the Indian Society of Remote Sensing*, 36(4), pp. 323–334.

Lancaster, B., 2006a. Guiding principles to welcome rain into your life and landscape. In: *Rainwater Harvesting for Drylands and Beyond*, vol. 1: Rainsource Press.

Lancaster, B., 2006b. Water-Harvesting Calculations [online]. Harvestingrainwater.com. Available at: https://www.harvestingrainwater.com/wp-content/uploads/Appendix3 Calculations.pdf; [Accessed 12 March 2020].

Li, L., Ren, J. and Zhu, Q., 2017. On The Application of Lora LPWAN Technology in Sailing Monitoring System [online]. 2017. wons-conference.org. Available at: https://2017. wons- conference.org/Papers/1570315122.pdf; [Accessed 12 March 2020].

Liaw, C.H., and Tsai, Y.L., 2004. Optimum Storage Volume of Rooftop Rainwater Harvesting Systems for Domestic Use. *Journal of the American Water Resources Association* 40, pp. 901–912

Met Office. 2020a. Climate Summaries [online]. Available at: https://www.metoffice.gov.uk/research/climate/maps-and-data/summaries/index; [Accessed 12 March 2020].

Met Office. 2020b. UK Temperature, Rainfall and Sunshine Time Series [online]. Available at: https://www.metoffice.gov.uk/research/climate/maps-and-data/uk-temperature-rain-fall-and- sunshine-time-series> [Accessed 12 March 2020].

Met Office. 2020c. Historic Station Data [online]. Available at: https://www.metoffice.gov.uk/research/climate/maps-and-data/historic-station-data; [Accessed 21 March 2020].

Miller, C., 2017. How To Calculate Runoff Coefficient [online]. Hunker. Available at: https://www.hunker.com/12001914/how-to-calculate-runoff-coefficient; [Accessed 12 March 2020].

Prima, E., Munifaha, S., Salam, R., Aziz, M. and Suryani, A., 2017. Automatic Water Tank Filling System Controlled Using Arduino TM Based Sensor for Home Application. *Procedia Engineering*, 170, pp. 373–377.

Rahman, S., Khan, M., Akib, S., Din, N., Biswas, S. and Shirazi, S., 2014. Sustainability of Rainwater Harvesting System in terms of Water Quality. *The Scientific World Journal*, 2014, pp. 1–10.

Reda, H., Daely, P., Kharel, J. and Shin, S., 2017. On the Application of IoT: Meteorological Information Display System Based on LoRa Wireless Communication. *IETE Technical Review*, 35(3), pp. 256–265.

Sanches Fernandes, L., Terêncio, D. and Pacheco, F., 2015. Rainwater harvesting systems for low demanding applications. *Science of the Total Environment*, 529, pp. 91–100.

Southern Water: Water for life, Water and wastewater services for Kent, Sussex, Hampshire and the Isle of Wight. 2020. Regional Rainfall - Southern Water: Water for Life, Water and Wastewater Services for Kent, Sussex, Hampshire and The Isle of Wight. [online] Available at: https://www.southernwater.co.uk/water-for-life/regional-rainfall; [Accessed 12 March 2020].

Susilo, G., 2018. Rainwater Harvesting as Alternative Source of Sanitation Water in Indonesian Urban Area (Case Study: Bandar Lampung City). *Rekayasa Sipil*, 12(1), pp. 15–21.

Texas A&M AgriLife Extension Service (2019). Rainwater Harvesting-Catchment Area. Available at: https://rainwaterharvesting.tamu.edu/catchment-area/; [Accessed 22 March 2020].

Ward, S., Memon, F.A., and Butler, D., 2010. Harvested Rainwater Quality: The Importance of Appropriate Design. *Water Science and Technology*, 61 (7), pp. 1707–1714.

Xu, W., Fletcher, T., Duncan, H., Bergmann, D., Breman, J. and Burns, M., 2018. Improving the Multi-Objective Performance of Rainwater Harvesting Systems Using Real-Time Control Technology. *Water*, 10(2), p.147.

4 Rainwater Harvesting in Jordan
Inherited Resilience and Adaptability

Mohammad Talafha, Rania F. Aburamadan, and Qais Hamarneh

4.1 INTRODUCTION

Extreme water scarcity is a major challenge in the Hashemite Kingdom of Jordan that threatens its social development, economic growth and environment. In addition to the limited water resources and issues like climate change impacts, high population growth rates and frequent influxes of refugees due to regional instability, water scarcity is further exacerbated in Jordan. In fact, Jordan has been classified as the second most water scarce country in the world (Beithou et al. 2022), where the per-person annual share of renewable water resources is less than $100\,m^3$ per person, which is well below the threshold of $500\,m^3$ per person, which defines severe water scarcity (Damkjaer and Taylor 2017). Jordan is a lower-middle income nation (The World Bank 2017) that is home to about 11,000,000 people, including 761,059 registered refugees (UNHCR 2022b). Situated in the heart of the Middle East, the country's climate zones range from Mediterranean to semi-arid to desert climate (The World Bank 2021a, b). In general, Jordan has dry and warm summers and wet and cool winters, with average annual precipitations ranging from less than 50 to up to 700 mm (Hammouri et al. 2022).

There are only three main rivers in Jordan: the Yarmouk River (on the Syrian border), the Jordan River (on the western border) and the Zarqa River (Figure 4.1, which also shows annual rainfall). Upstream diversions and over-pumping by Israel and Syria (Fanack's Water Editorial Team 2022) have made the water supply for the Yarmouk and Jordan rivers highly unreliable for Jordan. In fact, it has been reported that the Jordan River's flow in the lower region is now reduced to only 2% of its natural flow (Bromberg 2008). In addition to the regional conflict, the establishment of Israel's upstream diversions and withdrawals have led to this catastrophic situation on the Jordan River (Fanack's Water Editorial Team 2022; Bromberg 2008). Due to the scarcity of surface water, groundwater is heavily depended on to satisfy the water needs of the Kingdom and constitutes about 70% of the drinking water supply (IWMI 2020). This dependence on groundwater has led to over abstraction and negative water balance in 10 out of the 12 groundwater water basins in Jordan (UNICEF

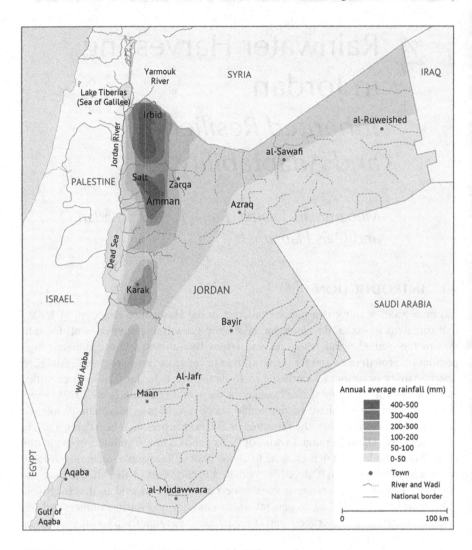

FIGURE 4.1   Annual rainfall in Jordan in 2015 (Source: fanack.com).

and Economist Impact 2022; Whitman 2019). The overexploitation of groundwater leads to detrimental effects, including water quality degradation, increased cost due to water table lowering, and land subsidence (The Groundwater Foundation 2022).

Despite the highly challenged water and economic situation, Jordan has managed to adapt and stay water resilient. According to the United Nation's SDG 6 data portal, WHO and UNICEF, 99% of the population use an improved drinking-water source, 86% of the population use safely managed drinking water services and 82% of the population use safely managed sanitation services (UN Water, 2020). Moreover, Jordan has a Sustainable Development Goals (SDG) index score of 67.4, which makes it the highest performing country in the Arab region (Bayoumi et al. 2022).

This work aims to report the story of the resilience of both ancient and modern Jordan in coping with limited water resources and water management adaptation through rainwater harvesting (RWH). Whether on the watershed or household scale. RWH is vital for the past, present and future of Jordan's water security, agriculture and economy. RWH is a green approach that diversifies water supply in a country that faces floods in winter yet droughts in summer. In Jordan and throughout history, it has been applied via different systems such as earth ponds, desert dams, cisterns, reservoirs, etc. as a lifeline that has enabled ancient civilizations to thrive for thousands of years in Jordan and is still part of the resilience and adaptability of the 21st century in Jordan. Examples of the history of these ancient practices in Jordan showcase the importance of RWH for survival. On the other hand, this work shows how modern Jordan can be an exemplar to other developing nations in terms of RWH policies, strategies, laws and technologies. Furthermore, challenges, opportunities and gaps in RWH in Jordan are discussed.

4.2 HISTORY OF RWH IN ANCIENT JORDAN

The locations of human settlements and the evolution of civilizations throughout history have always depended on the presence of a nearby water source. As an arid to semi-arid country with limited surface water resources, Jordan has faced major challenges in managing its water resources (Al-Kharabsheh 2020). Rainfall, the ultimate source of water, has always been the most important element for the people of the Jordanian lands. In summer, the significant increase in evaporation rates, exceeding 50%, poses challenges to maintaining water quality and sustainable consumption throughout the year (Al-Taani, Nazzal, and Howari 2021).

Therefore, managing the water resources and harvesting the greatest amount of rain to meet the water demands is one of the most important elements of the livelihood of the civilizations throughout Jordan's history. Ancient civilizations have managed to survive, settle, grow and prosper because they invented and implemented innovative water management systems that can provide insights into measures that may be deployed today to address Jordan's current water shortages. Jordan has a long history of water harvesting and water control systems (AbdelKhaleq and Alhaj Ahmed 2007), including dams, barrages, canals, qanat systems and reservoirs. Some of which are among the earliest examples in the world (Hammad 2022). This section will present an investigation of major water harvesting systems that were developed by different ancient civilizations throughout Jordan's history.

4.2.1 JAWA

The most ancient, well-researched and documented dams in the world to this day are the Jawa gravity dams in Jordan. The ancient city of Jawa (32.336 N, 37.002 E, 1002 m asl) is located in the basalt desert steppe of northeastern Jordan on the left bank of Wadi Rajil, about 7 km south of the present-day Syrian border (Figure 4.2). Jawa ruins date back to the Bronze Age, around 5,000 years ago (Mithen 2010). The climate in Jawa today is characterized by intense rainfall on a few days a year and dry weather with high temperatures for the rest of the year (AbdelKhaleq

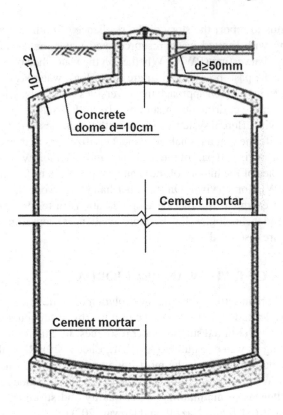

FIGURE 4.2   Jawa Photo by Robert Bewley Courtesy of APAAME (http://www.apaame.org).

and Alhaj Ahmed 2007). At the nearby Safawi climate station, mean annual rainfall totals about 70 mm/year and the annual temperature averages 19°C. Rainfall occurs mainly between November and March. The occurrence of strong winds is typical, and potential evaporation rates are high (Al-Homoud et al. 1995; Allison et al. 2000).

The people of Jawa developed a reliable water supply system by collecting runoff from winter precipitation and storing it in artificial cisterns (Hammad 2022). The system of Jawa depends mainly on large macro-watersheds and on several micro-watersheds. Wadi Rajil watershed was the city's main source of water, which posed a challenge for the people of Jawa to control such a discharge, which averages 2,000,000 m³ of water discharge each year (Al-Taani, Nazzal, and Howari 2021). Of this significant amount of water that runs off, only 3% (about 30,000 m³) flowed through the urban system (AbdelKhaleq and Alhaj Ahmed 2007). Water was stored at Jawa by deflection dams, which were constructed to direct parts of the water from Wadi Rajil to the canals leading to the storage areas.

The Jawa water system consists of three subsystems reflecting the three catchment areas (Al-Ansari, Al-Hanbaly, and Knutsson 2013). The first subsystem consists of a canal that conveys water from a deflection dam through a series of irrigable fields to an outflow gate, where part of the river empties into an underground reservoir.

The main canal continues to another drain, where it divides to fill three water reservoirs used for drinking water purposes. A branch of the main canal bypasses the storage area, runs through fields and reaches an animal watering place. All excess water returns to Wadi Rajil. Additional runoff is captured by three interconnected micro-catchments with deflection walls. The water from these is routed either to the bypass canal or to the three reservoirs (Helms 1981).

The second subsystem used a second deflection dam facing the eastern quarter of the lower town to raise water along the steep eastern shore of Wadi Rajil. A canal directed water to an outflow gate that fed two reservoirs (Al-Ansari, Al-Hanbaly, and Knutsson 2013). As in the first system, a micro-catchment area has been added. Runoff from the additional catchment flowed through a series of fields to a basin. The third subsystem was based on a deflection dam and channelled water through a gravity canal to several outflow barriers. The oldest and most complex urban water system yet discovered is the Jawa Water System (Helms 1981). Most of the old dams were simple gravity or earth-fill dams, but the Jawa Dam was constructed of gravel reinforced with rock fill behind the upstream wall to protect the wall from water pressure failure. This feature reflects the ancient designers' awareness of security measures that could be incorporated into the design to protect the structure (Adamo et al. 2020).

4.2.2 Petra

The ancient red rose city of Petra lies in the heart of the heights of southwestern Jordan with its outstanding rock art. The capital of the Nabatean Arabs is one of the most famous archaeological sites in the world, considered one of the seven New Wonders of the World. The Nabataean civilization originated in southern Jordan more than 2,500 years ago (Knodell et al. 2017). The Nabataeans excelled in science and engineering and were among the first to develop the sciences of astronomy, meteorology, navigation, geography and mathematics, in addition to their professionalism in architectural design and stone engraving of astonishing buildings, temples and tombs out of the sandstone layers that are exposed throughout the region (Al Farajat and Salameh 2010; Oleson 1992).

To survive and settle in this arid region, the Nabataeans had to develop and maintain advanced water management skills (Abdelal et al. 2021). They were pioneers in collecting, transporting and storing water, and they irrigated their land using an extensive and sophisticated system of dams, canals and reservoirs (Figure 4.3).

The water management of the Petra region can be divided into two main areas: (i) the supply of the city and (ii) the supply of the rural neighbourhoods (Ortloff 2005). The main sources of available water in the region are produced in the upper mountainous areas. The lower parts of the basin receive this water either as surface runoff (floodwater) or as groundwater at different times, mainly from springs located in limestone formations. Sandstone formations provide good conditions for harvesting surface runoff from precipitation events, especially in the lower parts (Abdelal et al. 2021). Predominantly agricultural terraces, cisterns and check dams—small, sometimes temporary structures constructed across swales, drainage ditches, or waterways—were built. These check dams serve the purpose of counteracting erosion by

FIGURE 4.3   Water open channel at Al Siq, Petra (photo by author Mohammad Talafha).

reducing water flow velocity. Accompanying these structures were respective water channels designed to supply water for rural neighbourhoods and agricultural purposes (Pfeil et al. 2018).

The seasonal stormwater runoff enters the valley through many canyons (wadis) and mainly drains through Wadi Siyagh. Petra's main water supply came from the Ain Mousa spring, about 7.0 km east of the town of Wadi Mousa, combined with water from the smaller Ain Umm Sar'ab spring, as illustrated in Figure 4.4. To access a more detailed water supply map and system, please refer to "Water Management in Petra" by Ortloff (2020). In the early stages of the Nabatean urban development, the main drinking water supply came from an open channel that carried the spring water from Ain Mousa through the Siq (Ortloff 2005). This channel extended through the urban core of the city to Q'asr al Bint with final drainage into Wadi Siyagh. In the

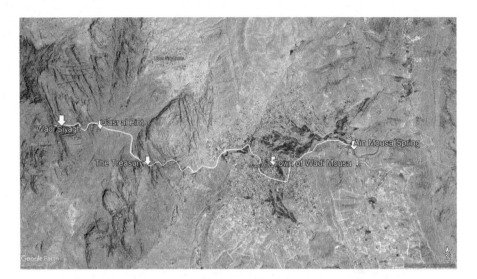

FIGURE 4.4   Petra's main water supply.

FIGURE 4.5   Nabataean water pipes (photo by author Mohammad Talafha).

time spanning, the first century BC to the first century AD, Nabataean builders substituted the initial Siq channel water supply system with a pipeline structure positioned along the northern wall of the Siq (Ortloff 2020). This pipeline was fed by a roughly 14-km conduit transporting water from the Ain Mousa spring. The sophisticated design of the piping system (Figure 4.5) in the rough terrain of the mountains and the hydraulic parameters of the system are clear evidence of the level of technical intelligence of the Nabatean engineers (Ortloff 2020).

What is unique is that the Nabataeans used every slope and surface as a means of harvesting rain and storing every source of water, from a few drops to large floods

FIGURE 4.6   Small cistern at the high place of sacrifice in Petra (photo by author Mohammad Talafha).

(Mays 2010). The Nabataeans used various types of cisterns, which they carved out of the rock and waterproofed with chalk. These cisterns ranged from small basins in the highlands to catch runoff, as shown in Figure 4.6, to rectangular-shaped cisterns at the bottom of the natural drips (Mays 2010).

The Khattara is an innovative system that collects humidity along a slope, saving water all year round as weak sludge is harvested from the walls because of the high-altitude condensation through drip tanks and cisterns. High water is diverted to water intakes and decanting basins to fill the large cistern where the water is stored. A staircase leads to the cistern, and water is drawn from the cistern through a well (Mays 2010). The reservoir at Zurraba (al Birka) is an example of later-phase technological advances. It was built to store and transport water along the Wadi Shab Qais around the north flank of Jebel el Khubtha Mountain in an elevated canal containing pipes that continued across royal tombs to feed a large basin at its end (Ortloff 2005).

To protect the city and the Siq from life-threatening floods, the Nabateans built an intricate hydraulic system that included a diversion dam and a 40-m tunnel at the entrance of ancient Petra. In the Siq, Dams were erected across smaller side wadis along the Siq to contain the water and create small reservoirs for use in the dry season (Figure 4.7). Field surveys have documented more than 35 dams and

FIGURE 4.7   Nabataean water dam (photo by author Mohammad Talafha).

cisterns, many channels and more than 400 wadi barriers and terraces within Petra city. The estimated total capacity of the dams and cisterns is 11,000 m³ of water (Akasheh 2002).

4.2.3   HUMAYMAH

Humaymah, the ancient site of Auara, was a major Nabatean settlement in the Hisma, the southern desert of Jordan. The site flourished during the Nabatean, Roman, Byzantine and Umayyad periods (Campana 2014). The site is a great example of several water harvesting techniques that have been used to store and supply rainwater. The average rainfall in the area is around 80 mm per year. Based on a survey conducted in 1986, 51 cisterns, four springs, one aqueduct, one dam, two sets of wadi barriers and six sets of terraces were identified in the targeted area. One year later, another survey of the water supply system within the settlement documented 16 structures: 11 cisterns, two reservoirs, two sets of pipes or drains and one bath building (Nydahl 2002).

Cisterns were either unroofed, roofed with stone slabs, or under undisturbed bedrock. Fourteen of which are roofed with transversal arches in a typical Nabataean manner, as shown in Figure 4.8. Two cisterns, No. 67 and No. 68, are great examples of Nabatean RWH technology and capability. They were built to harvest water from a large field covering about 100 hectares north of the settlement. Both are rectangular, narrow and roofed with stone slabs supported by 16 transverse arches. Cistern 67 has a 25 m long intake channel that was renovated in the 1960s. On the other hand, the

FIGURE 4.8   Reservoir 68 with water inside (photo by author Mohammad Talafha).

inlet channel of cistern 68 was constructed of large, heavy stone slabs, has a deep settling basin and was roofed with a slab supported by two transverse arches. Large parts of the roof are still supported by the remaining 14 arches (Oleson 1992).

The other nine cisterns in the centre of the settlement are all built of blocks. Since the settlement lies on the edge of fertile loessal plains, the bedrock is inaccessible. The lower capacity and the fact that they are usually partially buried in structural remains seem to indicate that they were intended for private use and built by individual families (Oleson 1992).

4.2.4   Umm El-Jimal

Umm El-Jimal is located north of Jordan, about 20 km east of Mafraq. It is primarily notable for the considerable ruins of a Byzantine and early Islamic city clearly visible above ground, as well as an older Roman village (locally referred to as al-Herri)

FIGURE 4.9   Rainwater harvesting at Umm el-Jimal (Photo by V.I. Taillefert).

located southwest of the Byzantine ruins. The city and its water system date back to 324-64 BC (UNESCO 2018). The water harvesting system consists of dams, canals and underground reservoirs (Figure 4.9). The dams are located in the higher parts of the mountains to collect and divert the water to the canals and then to the reservoirs. Rainfall events are short and intense, but the rate of evaporation is very high (Khoera 1990). These extreme conditions forced the people of the region to develop appropriate technologies to collect and store the water to be used.

Three dams were built in this area. They were built of basalt rock covered with white plaster mixed with gravel. They had no moving parts except for one that had something like a dump valve that was manually triggered when there was flooding in the area. The first dam was 1.7 m high and directly connected to the second dam, which is 0.8 m high. The second dam was a diversion dam capable of diverting 100,000 m³/hour, while the third dam was 32 m long and about 1.4 m high (Al-Ansari, Al-Hanbaly, and Knutsson 2013). A network of surface channels carrying water within the city walls was connected to the dams (Khoera 1990). The length of the main channel reaches 4 km. There are other side canals that were used to distribute water within the city. Basalt was used in the construction of the canals. The distribution of the canals was achieved by sluice gates. Thatched canals were used between the reservoirs within the city (Al-Ansari, Al-Hanbaly, and Knutsson 2013).

There are several open and covered reservoirs inside and outside the city of Umm El-Jimal (Oleson 1995). The total storage capacity of the reservoirs reaches more than 7,000 m³ (Evenari, Shanan, and Tadmor 1982). The water was stored in covered reservoirs to prevent evaporation and to protect the public and animals from falling in, as most of the reservoirs were within city limits. The Umm El-Jimal water harvesting systems were designed to minimize evaporative losses and energy to transport the harvested water (Al-Ansari, Al-Hanbaly, and Knutsson 2013).

4.2.5 GADARA AQUEDUCT

Beginning in the early or mid-second century AD, Roman water supply systems underwent evolutionary changes. The Gadara Aqueduct (The Qanaʾt Firʾaun, Decapolis Aqueduct and Yarmouk Aqueduct) was an aqueduct that served the Decapolis cities of Adraha (known today as Deraʾa in Syria), Abila (at Wadi Queilebh in Jordan) and Gadara (modern-day Umm Qais in Jordan) (Doering 2009). The total length of the aqueduct is more than 170 km. The aqueduct thus seems to be the most elaborately built long-distance pipeline of Roman times; The 106 km underground section is the longest of antiquity discovered to date (Doering 2009). Many parts are well preserved, but several sections are damaged by human intervention, such as road construction, or filled with sediment and soil. In many places, two parallel systems are recorded, one above the other. The dimension of the tunnel varies from area to area. It often reaches a height of about 3 m and a width of 1.5 m (Figure 4.10). The entrances to the tunnel are mostly inclined shafts, often equipped with stairs. It was probably the most important technological response to solving the region's water scarcity problem by minimizing evaporation and water loss through seepage. In addition, relying on gravity alone was energy efficient, eliminating the power-intensive need to pump water from the valleys to the cities on the plateaus (Abdel-Haleem Al-Malabeh and Kempe 2008).

Various archaeological sites have been excavated and artefacts recovered along the length of the tunnel, indicating its importance for settlement and agricultural activities in the past. Possibly, the system could be explored and unlocked to reactivate it for recent urbanization in the area. The use of filters through the tunnel walls could provide a new means of water harvesting, strengthening and replenishing local springs. Locally, many segments or entrances of this aqueduct have long been known. Currently, the water collected in the tunnel is used for various purposes (e.g., irrigation, horticulture and drinking water) (Abdel-Haleem Al-Malabeh and Kempe 2008).

The decline of the water harvesting systems developed by ancient civilizations in Jordan can be attributed to a multifaceted interplay of factors (Al-Addous et al. 2023). The lack of preservation efforts for these historic sites has left them vulnerable to degradation, particularly due to urbanization expanding above existing ancient water systems. The encroachment of modern developments has resulted in damage to these intricate water infrastructures. Additionally, geopolitical considerations have played a role, particularly in cases where water systems are transboundary, leading to closures and impacts on the sustainability of these ancient practices (Tayia 2019).

Climatic factors, including changes in rainfall patterns and prolonged droughts, have further strained the reliability of RWH systems. Furthermore, the abandonment of these ancient practices is also tied to the mismatch between the historical water

FIGURE 4.10   Flooded section of the Qanat Fir'aun. Image credit: Pafnutius – CC BY-SA 3.0.

systems and the rapidly growing demands of the modern population. The need for more advanced and efficient water systems for both drinking and agricultural purposes has contributed to the transition away from traditional RWH methods. As societies evolved, the pursuit of more sophisticated water infrastructure became essential to meet the increasing demands of a burgeoning population and ensure the sustainability of water resources in the region.

4.3 CURRENT RWH AND MANAGEMENT IN JORDAN

4.3.1 THE LEGAL FRAMEWORK

It is well established in Jordan that water and sustainable development are key national priorities, and strong commitment is displayed, especially by policymakers and water stakeholders (MWI 2016). Several national strategies, policies and action plans have been put in place by the government in Jordan to integrate the efficient management of water resources in Jordan, which are summarized in Figure 4.11.

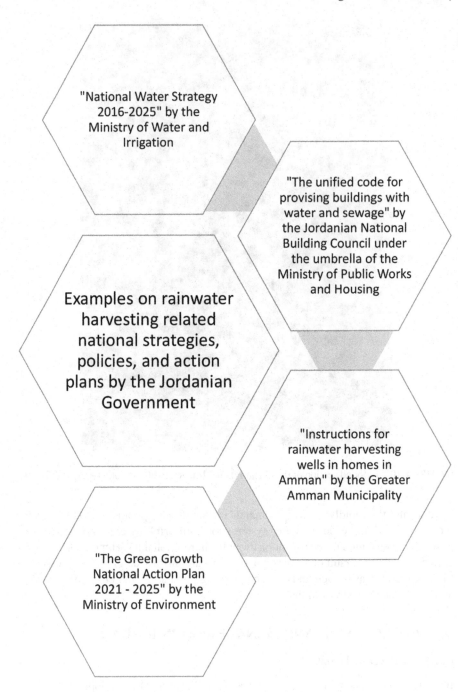

FIGURE 4.11   Legal Framework in Jordan governing rainwater harvesting.

4.3.1.1 National Water Strategy 2016–2025

In the national water strategy 2016–2025, the Jordanian Ministry of Water and Irrigation acknowledges that it is inevitable that the future development of water resources requires finding new sources of water and that RWH is one of the tools to make that happen (MWI 2016). As part of the planned development of water resources and under The Ministry of Water and Irrigation's (MWI) capital investment plan, the ministry is planning to implement household water harvesting projects that will increase the water supply by 7 million cubic metres, as well as desert and excavation dam water harvesting projects in remote areas that will supply an additional 15 million cubic metres of water.

In addition to promoting RWH as a means of increasing the water supply, MWI will be promoting the role of RWH as a means of improving irrigated agriculture and a component of the comprehensive risk management systems that guarantee soil fertility, agricultural labourers' health, as well as the hygiene of agricultural produce. Moreover, the strategy calls for collaborative partnerships that bring about new and alternative technologies like RWH to make water supply more efficient, cost-effective and sustainable.

The Green Growth National Action Plan 2021–2025 (GG-NAP) is a designed and strategic approach developed by the Jordanian Government that indicates actions and frameworks that implement sustainable and green growth in the sectors of Agriculture, Energy, Tourism, Transport, Waste and Water (Mo 2020). The GG-NAP's national green growth objectives are:

1. Enhance Nature Capital
2. Sustainable Economic Growth
3. Social Development and Poverty Reduction
4. Resource Efficiency
5. Climate Change Adaptation and Mitigation

From these five main green growth objectives, 16 water sector sub-objectives were identified, including "Augmenting water supply for priority economic activities through decentralized infrastructure solutions, such as RWH or reclaimed wastewater", as well as "Improve flood resilience through flood risk management measures, through appropriate flood mitigation infrastructure and measures to respond effectively to floods". To realize the 16 water sector sub-objectives, 19 water sector priority actions were identified in the water sector GG-NAP with five of them including the prospective implementation and increasing the practice of RWH in the kingdom (Table 4.1).

4.3.1.2 The Unified Code for Providing Buildings with Water and Sewage

Under the umbrella of the Ministry of Public Works and Housing, the Jordanian National Building Council set the unified code for providing buildings with water and sewage (Agricultural Lending Institution 2021). This code clearly defines the concept of water harvesting, its components and restrictions on the uses of water that is collected through the process of water harvesting, as well as the requirements for establishing a water harvesting system. The code determines the dimensions for

TABLE 4.1

Action Plans in Jordan that Aim to Expand RWH, Retrieved and Modified from Jordan's Water Sector GG-NAP 2021–2025 (MoE 2020)

	Estimated Cost (USD)	Implementation Milestones
WR02	Establish a rainwater harvesting project financing facility to support projects that argument rural and urban water supply	15,000,000	→ A pipeline of key small and medium water harvesting projects developed. → Funding mechanism/requirements determined and revenue model developed. → Structure and mandates of the responsible financing facility for managing RWH funds/projects identified. → Technical capacity of responsible body to manage RWH funds/projects ensured. → Government funds channelled into identified financing facility to support implementation of RWH projects. → Marketing and outreach plan developed and implemented to increase awareness. → RWH mainstreaming and expansion plan developed for the period 2025-2029. → Funding mechanism tested and piloted.
WR05	Construct dams and implement a parallel community water resource stewardship program for several communities in the Jordan Valley	66,900,000	→ Conduct strategic environmental impact assessments for 3 dams in the Jordan Valley. → Design and implement community-level climate change awareness and environmental stewardship programs. → Undertake a study to identify livelihood opportunities and innovative approaches to promote profitable, resource-efficient ventures. → Construction of dams.
WR12	Increase the resilience of displaced persons (DP) and host communities to climate change related water and climate change challenges	7,000,000	→ Land use strategies and plans. → Capacity building programs of municipal officers conducted. → Participatory community-level planning processes developed and implemented to promote social exchange focused on water conservation and adaptation options to climate change. → Community level skill building trainings conducted. → Plans for water conservation and climate adaptation developed and implemented. → Regional' urban risks and vulnerabilities assessment, planning and management approach model developed.

(Continued)

TABLE 4.1 (*Continued*)

Action Plans in Jordan that Aim to Expand RWH, Retrieved and Modified from Jordan's Water Sector GG-NAP 2021–2025 (MoE 2020)

		Estimated Cost (USD)	Implementation Milestones
WR13	Increase the availability of WASH in Schools and strengthening WinS (WASH in Schools) standards for climate change impacts	1,500,000	→ Target schools develop and implement local water conservation plans. → Target schools develop and implement wastewater reuse strategies. → Target schools can recognize and mitigate water insecurity drivers within their campuses. → Increased awareness of the benefits of water conservation and incentives to practice water conservation behaviours. → Water conservation and wastewater reuse strategy incorporated. → Benefits of installing Decentralized Wastewater Treatment Systems (DEWATS) and Rain Harvesting technologies assessed. → A comprehensive complementary package developed, focusing on the promotion of green growth.
WR18	Undertake feasibility studies to explore storm water systems and groundwater filtration	15,000,000	→ Hydrological assessments and climate risk assessments conducted to inform climate change adaptation solutions for flood management in target cities. → A diagnostic will be led and recommendations will be issued as to structural and nonstructural measures that can be implemented locally by residents, businesses, or public institutions to reduce flood vulnerability. → Develop a comprehensive GIS on flood risk. → Develop a storm water drainage master plan for target cities. → Design and install LID/EbA solutions in target cities.

rainwater collection tanks based on factors such as the construction site's location and scale. Nonetheless, it is imperative to revise the National Building Code to incorporate water-saving systems (Abdulla 2020).

4.3.1.3 The Instructions for RWH Wells in Homes in Amman

The process of establishing an RWH system became a requirement when requesting a building permit for any new building within the Greater Amman Municipality. According to the legislative hierarchy in Jordan in 2020, the instructions for RWH wells in homes in Amman were issued in accordance with the article (49) of the buildings and organization in the city of Amman regulation number (28) of the year 2018, which was issued in accordance with the organizing law of cities, towns and villages. According to the instructions, the minimum volume of an RWH well depends on the following factors:

1. The average annual precipitation rate in mm and its distribution. The instructions clearly identify the distribution of the average annual precipitation rate in the areas of Amman (Table 4.2).

TABLE 4.2
Distribution of Mean Annual Precipitation Rates in Amman (USAID 2018)

Number	Area	Annual precipitation Rate (mm)
1	Al Madina	341
2	Basman	256
3	Marka	168
4	Al Naser	338
5	Al Yarmouk	312
6	Ras Al Ain	312
7	Badr	404
8	Zahran	437
9	Al Abdali	297
10	Tareq	226
11	Al Quwaismeh	283
12	Khraibet Al Souq	277
13	Al Muqabalain	330
14	Wadi Al Seer	391
15	Badr Al Jadidah	369
16	Suwaileh	400
17	Tila' Al Ali	368
18	Al Jubaihah	332
19	Shafa Badran	332
20	Abu Nussair	393
21	Ohod	170
22	Marj Al Hamam	336

2. The surface area, which is the area of the RWH surface's horizontal projection in metres squared.
3. The runoff coefficient, which is equal to 0.8 according to the Jordanian unified code for providing buildings with water and sewage.

The instructions also provide the equations that can be used to calculate the volume of RWH:

For areas with an average annual precipitation rate less than 500 mm: volume = annual precipitation rate (mm) × surface area (m²) × runoff coefficient × 0.46

For areas with an average precipitation rate equal to or more than 500 mm: volume = annual precipitation rate (mm) × surface area (m²) × runoff coefficient × 0.44

4.4 CURRENT RWH INFRASTRUCTURE AND PRACTICES IN JORDAN

4.4.1 STORMWATER AND WATERSHED SCALE RWH IN JORDAN

In 2019, 30.8% of the supplied water in Jordan was from surface water (wadis, springs and run-off), 54.4% from groundwater, 14.5% treated wastewater and 0.3% desalinated seawater (MWI 2019a). Since 1967 (Alsilawi 2021), The Ministry of Water and Irrigation (MWI) has been constructing several RWH dams to secure water for potable, agricultural, industrial and touristic uses (Table 4.3).

The Jordanian eastern region, also known as the Badia, comprises about 75% of Jordan's area and is part of the Syrian Desert, which receives significant amounts of floods resulting from torrential rains during the wet season. Currently, the Badia in Jordan is home to 330 earth ponds and desert dams (MWI 2019a). An earth pond serves as an alternative RWH technique that is decentralized, simple and requires simple machinery to construct (Alkhaddar 2003). The harvested rainwater from earth ponds and desert dams can be used for groundwater recharge, agriculture, pasture improvement and watering animals. In October 2013, the capacity of existing water harvesting structures (dams, ponds and pools) was 110,755 MCM (FAO 2016).

4.4.2 HOUSEHOLD SCALE RWH

Decentralized household RWH is a common and traditional practice in Jordan, especially in the north-western region, where rainwater is collected from rooftops and stored. The two most common rainwater storage systems in Jordan are pear-shaped underground wells (commonly known as "injassa wells") and reinforced concrete tanks, while some areas use plastic tanks (Assayed and Hazaymeh 2018). In the guideline for greywater treatment and reuse and RWH and collection at home (Assayed and Hazaymeh 2018), it has been estimated that, each year, a total of 28.7 million cubic meters (MCM) of rainwater can be harvested from the roofs of Jordanian households. When compared to the annual water use in Jordan, which is

TABLE 4.3
Main Water Dams in Jordan (MWI 2019a, b)

Dam	Potable	Irrigation	Industry	Hydropower	GW Recharge	Capacity (MCM)
			Purpose			
Wehdeh	X	X				110
King Talal		X		X		86
Karameh	X* desalination	X				55
Mujeb	X	X	X			32
Wadi Arab	X	X	X	X		20
Tanour		X				16.8
Kafrain		X			X	8.5
Wala		X			X	8.2* to be increased to 25
Kufranjeh	X	X			X	7.8
Zeqlab		X				4.3
Wadi Shueib		X			X	2.3
Zarqa Main	X					2
Karak	X		X			2
Lajoon		X			X	1

1053.6 MCM (MWI 2017), this amount of harvested rainwater can provide about 3% of Jordan's total water use needs or about 6% of Jordan's domestic water use needs (469.7 MCM).

4.5 EFFORTS TO SCALE-UP RAINWATER HARVESTING

All stakeholders in Jordan, whether local or international, are making diligent efforts to scale up rainwater harvesting in Jordan through different initiatives and projects. The following are some examples of such efforts:

> To scale up rainwater harvesting to support the agricultural sector, the Agricultural Credit Corporation of Jordan allocated circa 14.09 USD for its interest-free small loans programme to fund innovative agricultural projects, including rainwater harvesting systems (Agricultural Lending Institution 2021).
> Through several community-based revolving funds, more than 3000 RWH cisterns were implemented by the United Nations for the local community (Assayed and Hazaymeh 2018).
> Through the 5-year (2006–2011) project "Community Based Initiatives for Water Demand Management" implemented by Mercy Corps with funding

from the USAID, more than 1,777 on-site rainwater harvesting cisterns were implemented with an annual water harvesting capacity of $88,335\,m^3$ (Assayed et al. 2013). This project was able to meet about 24% of the water needs of benefited households.

In 2021, the Green Climate Fund granted Jordan a 25 million USD grant to build climate resilience through better water management practices (Al Khayyat 2021). With additional co-funding from the Jordanian Government and United Nations agencies, the Food and Agriculture Organization (FAO) will be implementing this project, including water harvesting wells, to benefit about 250,000 people in four targeted areas.

4.6 OPPORTUNITIES AND CHALLENGES

Jordan has reached several milestones in water management; nonetheless, there are still many challenges and barriers to overcome with water management approaches like RWH to support expansion and scaling up on all levels.

Climate-related hazards such as flash floods hit Jordan frequently (Mo 2021). However, due to the scarcity situation, the country needs every drop of water. Most Jordanians still do not have a constant supply of water, but rather they depend on storage tanks to store their intermitted water supply, which ranges from twice a week to once every 2 weeks (Pawson 2021). The UNHCR ranked Jordan the number two country worldwide in the number of Syrian refugees per capita (UNHCR 2022a). Refugees and displaced people (DP), especially in the Northern and Eastern regions of Jordan, are also highly vulnerable to water challenges. A survey by the Jordan Response Plan (JRP) showed that Syrian DPs were spending over 25% of their expenditure on water, sanitation and hygiene (MoPIC 2015). Almost each winter, urban areas, including the capital Amman suffer from floods during heavy rainstorms that result in infrastructure damage as well as financial and human losses (Al Nawas 2020; Arab News 2019).

Despite the challenges, it is evident that storm and rainwater harvesting stands as an innovative and decentralized approach and opportunity to conserve water, mitigate climate change impacts and increase the resilience of Jordanian cities, rural communities, students and displaced people.

Blue-Green Infrastructures (BGI) are considered integrated innovative solutions to improve water management and landscape values for more climate-resilient and liveable cities. BGI creates the opportunity to renew the natural structure of the water balance in cities by increasing rainwater retention and enlarging permeability areas (Pochodyła, Glińska-Lewczuk, and Jaszczak 2021). For Jordan, there are a variety of BGI solutions that benefit the urban environment, including, but not limited to, permeable pavement, green ditches/grassed bioswales, bioretention cells, infiltration trenches and urban trees. A holistic, context-specific and participatory approach should be considered not only for the urban context but also for the rural communities in the Badia region and the villages of Jordan.

In 2016 (yet to be updated), the Food and Agriculture Organization of the United Nations (FAO)'s assessment of the water harvesting sector in Jordan identified the existing, on-going and future water harvesting sites in collaboration with the Ministry of Agriculture and the Jordan Valley Authority (FAO 2016) (Figure 4.12).

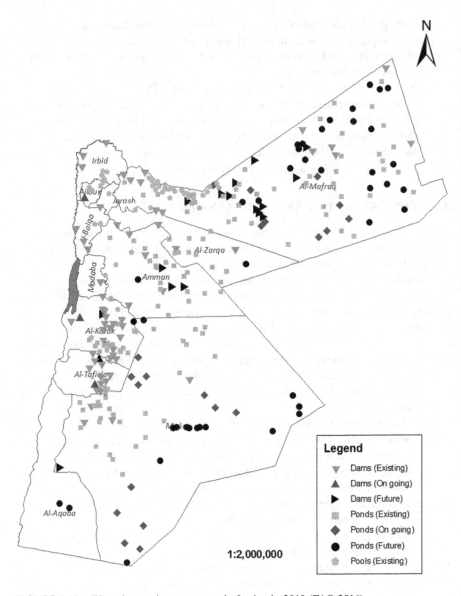

FIGURE 4.12 Water harvesting structures in Jordan in 2013 (FAO 2016).

The water harvesting sector in Jordan still faces some multi-sectoral gaps (FAO 2018), some of which were identified in a 2018 assessment of policies, institutions and regulations as well as the GG-NAP:

Administrative gap: Despite having a clear domain by the Water Harvesting Directorate and Ministry of Agriculture (MoA) on the coordination and planning of water harvesting structures, it is yet unclear in Jordan who administers the operation and maintenance of such structures. Therefore, it is recommended to establish and authorize a department responsible for supervising the operation and maintenance of water harvesting structures. The administrative gap should be tackled in parallel with the following recommendations for the information and planning gap.

Information and planning gap: There is still no database for planning and monitoring water harvesting structures. The inexistence of such a database has led to a lack of framework cooperation between stakeholders, individual case planning, and water harvesting measure implementation being distant from communities and agriculture reuse sites. Therefore, it is recommended to develop a comprehensive geodatabase including all physical, geographic, hydrologic, social and other aspects of water harvesting planning. A similar database for sanitation planning was developed by the Bremen Overseas Research and Development Association (BORDA) in collaboration with MWI and the Water Authority of Jordan (WAJ) (Hammouri et al. 2022). Moreover, this database should include all previously performed feasibility and technical assessments. The database should also be the foundation of a new framework of coordination between all acting stakeholders that engages all relevant authorities, donor entities, NGOs and CBOs and rural communities and farmers. It is recommended that this framework should, especially, specify an administrative matrix for all the phases of water harvesting structures.

Funding gap: The Water Sector GG-NAP 2021-2025 has listed ambitious action plans to scale up RWH in the kingdom and identified the needed budget. Nonetheless, the funding for such action plans is still not fully secured. Moreover, communities and farmers living in Badia and rural areas do not have the financial means to implement household or community-based RWH systems. It is therefore recommended to increase budget allocations to RWH projects, encourage investors and donors to fund RWH implementations as per the national strategies and action plans, and develop financial incentives for communities and farmers. According to the Royal Scientific Society, there is no economic feasibility or cost-benefit analysis for RWH on a household scale that considers the direct and indirect costs of RWH wells or cisterns (Assayed and Hazaymeh 2018). An economic feasibility and cost-benefit analysis for household-scale RWH systems is highly recommended, where the analysis and benefit of RWH systems shall consider the direct factors like the cost of raw materials and size of tanks, as well as the indirect benefits such as the reduced cost of desludging, social conflicts and competition, increased real estate value and the reduction of groundwater exploitation.

Capacity gap: Whether on an institutional or community level, the capacity of qualified persons to plan, implement, operate, maintain and manage RWH systems is still lacking. In addition, the holistic knowledge of linking RWH to multi-aspects such as climate change, new technologies and tools, economic benefit, humanitarian side and agriculture is still not efficiently understood nor integrated. It is therefore recommended that capacity building and awareness raising be implemented through technical and

non-technical training programmes, workshops, media, educational curri-
cula for institutions, local professionals, communities, farmers, youth, students and refugees.

Inclusive involvement gap: Going back to the previously discussed legal framework and action plans, one can notice an evident gap in the involvement
of different local communities' involvement and customary laws. As each
local community in each region in Jordan may require a different decentralized rainwater harvesting approach or customary laws based on the climatic
and geographic characteristics of their region, it is recommended to involve
and engage local community representatives of different regions through
focus groups or workshops in the process of policy and strategy making.

4.7   CONCLUSION

This chapter extensively explores Jordan's adept management of its limited water
resources throughout time, notably highlighting the significant role played by rainwater harvesting. Through a thorough examination spanning historical and contemporary periods, the chapter underscores Jordan's remarkable adaptability in water
management. The historical insights into ancient rainwater harvesting technologies
in Jordan vividly emphasize the enduring importance of this practice for the survival
of settlements and communities. This collective narrative highlights rainwater harvesting as a fundamental element of Jordan's resilience in addressing water scarcity
challenges over time. Jordan is on track with its progress in water management, rec-
ognizing the significance of alternative water sources like rainwater harvesting, especially given its arid climate and scarcity of water resources. The demand for meeting
the needs of a growing population and hosting refugees across several Jordanian cities
has significantly strained the country's water management strategies. Climate-related
issues further compound the challenges due to Jordan's arid climate. In this context,
rainwater harvesting emerges as a critical approach to address climate challenges and
water scarcity issues by leveraging variations in arid weather patterns. Various initiatives promoting rainwater harvesting have showcased its potential to complement
traditional water sources. However, challenges persist and remain, particularly in the
widespread adoption and implementation of rainwater harvesting approaches across
all levels within the country. Introducing rainwater harvesting as a regulatory measure
might face challenges in the capital city compared to more remote areas organized
under a decentralization system. This disparity in implementing rainwater harvesting
strategies between urban centres and decentralized regions could pose a potential
hurdle to its successful integration. Despite these ongoing efforts, barriers to the wider
application of rainwater harvesting practices in Jordan remain, and various gaps are
still faced, including administrative, information and planning, funding, capacity and
inclusive involvement gaps. Recommendations include establishing a department for
the operation and maintenance of water harvesting structures, creating a geodatabase
for planning and monitoring, securing funding for RWH projects, enhancing capacity through training programmes and involving local communities in policy-making
processes. Addressing these gaps is crucial for the successful implementation and
sustainability of rainwater harvesting initiatives in Jordan.

REFERENCES

Abdelal, Qasem, Abdulla Al-Rawabdeh, Khaldoon Al Qudah, Catreena Hamarneh, and Nizar Abu-Jaber. 2021. "Hydrological Assessment and Management Implications for the Ancient Nabataean Flood Control System in Petra, Jordan." *Journal of Hydrology* 601 (October): 126583. https://doi.org/10.1016/j.jhydrol.2021.126583.

Abdel-Haleem Al-Malabeh, Ahmad, and Stephan Kempe. 2008. "The Decapolis Aqueduct Tunnel System (Al-Tura-Umm Quis, Jordan): The Longest in Antique History." https://www.researchgate.net/publication/256722370.

AbdelKhaleq, R.A., and I. Alhaj Ahmed. 2007. "Rainwater Harvesting in Ancient Civilizations in Jordan." *Water Supply* 7 (1): 85–93. https://doi.org/10.2166/ws.2007.010.

Abdulla, Fayez. 2020. "Rainwater Harvesting in Jordan: Potential Water Saving, Optimal Tank Sizing and Economic Analysis." *Urban Water Journal* 17 (5): 446–456. https://doi.org/10.1080/1573062X.2019.1648530.

Adamo, Nasrat, Nadhir Al-Ansari, Varoujan Sissakian, Jan Laue, and Sven Knutsson. 2020. "Dam Safety: General Considerations." *Journal of Earth Sciences and Geotechnical Engineering* 10: 1–20.

Agricultural Lending Institution. 2021. "وثيقة حزمة المشاريع لدعم القطاع الزراعي للعام (2021) (بدون" (فائده." https:// www.Acc.Gov.Jo/Ar/Node/559. April 11, 2021.

Akasheh, Talal S. 2002. "Ancient and Modern Watershed Management in Petra." *Near Eastern Archaeology* 65 (4): 2002.

Al-Addous, Mohammad, Mathhar Bdour, Mohammad Alnaief, Shatha Rabaiah, and Norman Schweimanns. 2023. "Water Resources in Jordan: A Review of Current Challenges and Future Opportunities." *Water* 15 (21): 3729. https://doi.org/10.3390/w15213729.

Al-Ansari, Nadhir, Mariam Al-Hanbaly, and Sven Knutsson. 2013. "Hydrology of the Most Ancient Water Harvesting Schemes." *Journal of Earth Sciences and Geotechnical Engineering* 3 (1): 15–25.

Al Farajat, Mohammad, and Elias Mechael Salameh. 2010. "Vulnerability of the Drinking Water Resources of the Nabataean of Petra." https://www.researchgate.net/publication/259177727.

Al-Homoud, Azm S., Robert J. Allison, Basam F. Sunna, and K. White. 1995. "Geology, Geomorphology, Hydrology, Groundwater and Physical Resources of the Desertified Badia Environment in Jordan." *The Muslim World* 37 (1): 51–67.

Alkhaddar, Rafid. 2003. "Water Harvesting in Jordan Using Earth Ponds." *Waterlines* 22 (2): 19–21.

Al-Kharabsheh, Atef. 2020. "Challenges to Sustainable Water Management in Jordan." *Jordan Journal of Earth and Environmental Sciences* 11 (1): 38–48.

Al Khayyat, Dima. 2021. "Jordan Receives a $25 Million Grant to Enhance Climate Change Adaptation." [Press Release], FAO, https://Jordan.Un.Org/En/122751-Jordan-Receives-25-Million-Grant-Enhance-Climate-Change-Adaptation. March 21, 2021.

Allison, Robert J., James R. Grove, David L. Higgitt, Alastair J. Kirk, Nicholas J. Rosser, and Jeff Warburton. 2000. "Geomorphology of the Eastern Badia Basalt Plateau, Jordan." *The Geographical Journal* 166 (4): 352–370. https://doi.org/10.1111/j.1475-4959.2000.tb00036.x.

Al Nawas, Bahaa Al Deen. 2020. "Over 100 Amman Shops Claim Damage after Recent Flooding." https://www.Jordantimes.Com/News/Local/over-100-Amman-Shops-Claim-Damage-after-Recent-Flooding, January 11, 2020.

Alsilawi, Rana. 2021. "What Are the Water Dams in Jordan?" https:// www.Arabiaweather.Com/En/Content/What-Are-the-Water-Dams-in-Jordan#:~:Text=Sharhabil%20bin%20Hasna%20Dam%20(Zaqlab,Is%20used%20for%20irrigation%20purposes. November 9, 2021.

Al-Taani, Ahmed A., Yousef Nazzal, and Fares M. Howari. 2021. "Groundwater Scarcity in the Middle East." In *Global Groundwater*, 163–175. Elsevier. https://doi.org/10.1016/B978-0-12-818172-0.00012-8.

Arab News. 2019. "Heavy Rainfall and Flooding Sparks Chaos in Amman." https:// www. Arabnews.Com/Node/1459456/Middle-East. February 28, 2019.

Assayed, Almoayied, and Ayat Hazaymeh. 2018. *Guideline for Greywater Treatment and Reuse and Rainwater Harvesting and Collection at Home- Arabic Version*. Amman, Jordan: Royal Scientific Society.

Assayed, Almoayied, Zaid Hatokay, Rania Al-Zoubi, Shadi Azzam, Mohammad Qbailat, Ahmad Al-Ulayyan, Ma'ab Abu Saleem, Shadi Bushnaq, and Robert Maroni. 2013. "On-Site Rainwater Harvesting to Achieve Household Water Security among Rural and Peri-Urban Communities in Jordan." *Resources, Conservation and Recycling* 73 (April): 72–77. https://doi.org/10.1016/j.resconrec.2013.01.010.

Bayoumi, Moustafa, Mari Luomi, Grayson Fuller, Aisha Al-Sarihi, Fadi Salem, and Seppe Verheyen. 2022. "2022 Arab Region SDG Index and Dashboard Report." www.Arab SDGIndex.com.

Beithou, Nabil, Ahmed Qandil, Mohammad Bani Khalid, Jelena Horvatinec, and Gabrijel Ondrasek. 2022. "Review of Agricultural-Related Water Security in Water-Scarce Countries: Jordan Case Study." *Agronomy* 12 (7): 1643. https://doi.org/10.3390/agronomy12071643.

Bromberg, Gidon. 2008. "Will the Jordan River Keep on Flowing?" https://E360.Yale.Edu/Features/Will_the_jordan_river_keep_on_flowing. September 18, 2008.

Campana, Ivan. 2014. "Water Supply and Hydraulic Devices: The Dams in the Umayyad Jordan." In *Proceedings - Irrigation, Society and Landscape. Tribute to Thomas F. Glick*, 1–12. Editorial Universitat Politècnica de València. https://doi.org/10.4995/ISL2014.2014.149.

Damkjaer, Simon, and Richard Taylor. 2017. "The Measurement of Water Scarcity: Defining a Meaningful Indicator." *Ambio* 46 (5): 513–531. https://doi.org/10.1007/s13280-017-0912-z.

Doering, Mathias. 2009. "Qanat FIR'ØN - Documentation of the 100 Kilometres Aqueduct Tunnel in Northern Jordan." University of Applied Science, Haardtring 100, D-64295 Darmstadt, Germany.

Evenari, Michael, Leslie Shanan, and Naphtali Tadmor. 1982. "Recent Developments." In *The Negev*, 338–412. Harvard University Press. https://doi.org/10.4159/harvard.9780674419254.c20.

Fanack's Water Editorial Team. 2022. "Country Report: Jordan Water Report." June 2, 2022.

Salman, Maher, Motasem Abu Khalaf, Brenna Moore, Ahmad Al Qawabah, and Ayman Al Hadid. 2016. "Assessment of the Water Harvesting Sector in Jordan." Rome: Food and Agriculture Organization of the United Nations (FAO). https://openknowledge.fao.org/server/api/core/bitstreams/7a22f3c6-8a30-411e-9182-ab0496601cda/content.

Salman, Maher, Claudia Casarotto, Maria Bucciarelli, and Maria Losacco. 2018. "An Assessment of Policies, Institutions and Regulations for Water Harvesting, Solar Energy, and Groundwater in Jordan." Rome: Food and Agriculture Organization of the United Nations (FAO), Land and Water Division (CBL). https://openknowledge.fao.org/server/api/core/bitstreams/43442867-1d8b-463c-b6b9-d90f5916b924/content.

Bromberg, Gidon. 2008. "Will the Jordan River Keep on Flowing?"

Hammad, Rizeq. 2022. "Water Harvesting in Archeological Buildings in Jordan." *Journal of Biosensors and Renewable Sources* 1 (5): 135–141. https://lupinepublishers.com/biosensors-renewable-sources/pdf/JBRS.MS.ID.000124.pdfhttps://doi.org/10.32474/JBRS.2022.01.000124.

Hammouri, Nezar, Mohammad Talafha, Qais Hamarneh, Zeina Annab, Rami Al-Ruzouq, and Abdallah Shanableh. 2022. "Vulnerability Hotspots Mapping for Enhancing Sanitation Services Provision: A Case Study of Jordan." *Water* 14 (11): 1689. https://doi.org/10.3390/w14111689.

Helms, Svend W. 1981. *Jawa, Lost City of the Black Desert*. 1st ed. Ithaca, NY: Cornell University Press.

IWMI. 2020. "Animated Groundwater Model of Jordan." https://Gripp.Iwmi.Org/2020/12/01/ Animated-Groundwater-Model-of-Jordan/#:~:Text=The%20country%20is%20heavily% 20dependent,Some%20of%20these%20are%20transboundary, December 1, 2020.

Khoera, A.M.Y. 1990. "The Irrigation System in Umm El-Jimal." Irbid, Jordan: Yarmouk University.

Knodell, Alex R., Susan E. Alcock, Christopher A. Tuttle, Christian F. Cloke, Tali Erickson-Gini, Cecelia Feldman, Gary O. Rollefson, Micaela Sinibaldi, Thomas M. Urban, and Clive Vella. 2017. "The Brown University Petra Archaeological Project: Landscape Archaeology in the Northern Hinterland of Petra, Jordan." *American Journal of Archaeology* 121 (4): 621–683. https://doi.org/10.3764/aja.121.4.0621.

Mays, L. 2010. *Ancient Water Technologies*. edited by L. Mays. Dordrecht: Springer Netherlands. https://doi.org/10.1007/978-90-481-8632-7.

Mithen, Steven. 2010. "The Domestication of Water: Water Management in the Ancient World and Its Prehistoric Origins in the Jordan Valley." *Philosophical Transactions of the Royal Society A: Mathematical, Physical and Engineering Sciences* 368 (1931): 5249–5274. https://doi.org/10.1098/rsta.2010.0191.

Mo E. 2020. "Water Sector Green Growth National Action Plan 2021 - 2025." Amman, Jordan, MoEnv. Ministry of Environment. https://www.moenv.gov.jo/ebv4.0/root_storage/ar/ eb_list_page/final_draft_nap-2021.pdf

Mo E. 2021. "The National Climate Change Adaptation Plan of Jordan." Amman, Jordan.

MoPIC. 2015. "The Jordan Response Plan (JRP)." Amman, Jordan. The Ministry of Planning and International Cooperationhttps://www.mop.gov.jo/EBV4.0/Root_Storage/AR/EB_List_ Page/JRP_2023_Narrative_Summary.pdf

MWI. 2016. "National Water Strategy of Jordan, 2016 - 2025." Amman, Jordan. The Ministry of Water and Irrigation (MWI).

MWI. 2017. "Jordan Water Sector Facts and Figures 2017." Amman, Jordan. The *Ministry of Water* and Irrigation (MWI).

MWI. 2019a. "Water Annual Report 2019." Amman, Jordan. The *Ministry of Water* and Irrigation (MWI).

MWI. 2019b. "Water Budget 2019 - Arabic Version." Amman, Jordan. The *Ministry of Water* and Irrigation (MWI).

Nydahl, Hanna. 2002. "Archaeology and Water Management in Jordan." Hanna Nydahl, Uppsala, Sweden. https://scholar.google.com/citations?view_op=view_citation&hl=sv &user=B-7ww7UAAAAJ&citation_for_view=B-7ww7UAAAAJ:u5HHmVD_uO8C

Oleson, John Peter. 1992. "The Origins and Design of Nabataean Water-Supply Systems." *Studies in the History and Archaeology of Jordan. Department of Antiquities* 4: 269–275.

Oleson, John Peter. 1995. "The Origins and Design of Nabataean Water-Supply Systems." *Studies in the History and Archaeology of Jordan. Department of Antiquities* 5: 707–719.

Ortloff, Charles R. 2005. "The Water Supply and Distribution System of the Nabataean City of Petra (Jordan), 300 BC–AC 300." *Cambridge Archaeological Journal* 15 (1): 93–109. https://doi.org/10.1017/S0959774305000053.

Ortloff, Charles R. 2020. "Hydraulic Engineering at 100 BC–AD 300 Nabataean Petra (Jordan)." *Water* 12 (12): 3498. https://doi.org/10.3390/w12123498.

Pawson, Melissa. 2021. "'Catastrophe' Faces Jordan's Water Sector as Climate Heats up." https://www.Aljazeera.Com/News/2021/11/2/Experts-Warn-of-Catastrophe-Facing-Jordans-Water-Sector#:~:Text=A%20recent%20study%20published%20in,The%20 end%20of%20the%20century. November 2, 2021.

Pfeil, Friedrich, München, Jonas Berking, and Brigitta Schütt. 2018. "Water Harvesting in Drylands Water Knowledge from the Past for Our Present and Future." Verlag Dr. Friedrich Pfeil, 2018. https://www.researchgate.net/publication/329585393_Water_Harvesting_ in_Drylands_-_Water_Knowledge_from_the_Past_for_our_Present_and_Future

Pochodyła, Ewelina, Katarzyna Glińska-Lewczuk, and Agnieszka Jaszczak. 2021. "Blue-Green Infrastructure as a New Trend and an Effective Tool for Water Management in Urban Areas." *Landscape Online* 92 (September): 1–20. https://doi.org/10.3097/LO.202192.

The Groundwater Foundation. 2022. "Groundwater Overuse and Depletion." https://Groundwater.Org/Threats/Overuse-Depletion/. 2022.

Tayia. 2019. "Transboundary Water Conflict Resolution Mechanisms: Substitutes or Complements." *Water* 11 (7): 1337. https://doi.org/10.3390/w11071337.

The World Bank. 2017. "Jordan Country Reclassification - Questions and Answers." https://www.Worldbank.Org/En/Country/Jordan/Brief/Qa-Jordan-Country-Reclassification#: ~:Text=On%20July%201%2C%202016%2C%20the,A%20lower%20middle%20 income%20country. July 6, 2017.

The World Bank. 2021a. "Agriculture, Forestry, and Fishing, Value Added (% of GDP) - Jordan." World Bank National Accounts Data, and OECD National Accounts Data Files. License : CC BY-4.0 . 2021.

The World Bank. 2021b. "Climatology." Climate Change Knowledge Portal. 2021.

UNESCO. 2018. "Umm Al-Jimal." https://Whc.Unesco.Org/En/Tentativelists/6335/. March 29, 2018.

UNHCR. 2022a. "Jordan Issues Record Number of Work Permits to Syrian Refugees." https://www.Unhcr.Org/News/News-Releases/Jordan-Issues-Record-Number-Work-Permits-Syrian-Refugees#:~:Text=Jordan%20hosts%20760%2C000%20refugees%20 and,Per%20capita%20globally%20behind%20Lebanon. January 25, 2022.

UNHCR. 2022b. "Jordan: UNHCR Operational Update". November 2022.

UNICEF and Economist Impact. 2022. "The Costs of Water Stress in Jordan."

USAID. 2018. "حصاد مياه الأمطار للمنازل." Amman, Jordan. https://www.mwi.gov.jo/ebv4.0/root_ storage/ar/eb_list_page/%D8%AD%D8%B5%D8%A7%D8%AF_%D9%85%D9%8A %D8%A7%D9%87_%D8%A7%D9%84%D8%A3%D9%85%D8%B7%D8%A7%D8 %B1_%D9%84%D9%84%D9%85%D9%86%D8%A7%D8%B2%D9%84_2018.pdf.

Whitman, Elizabeth. 2019. "A Land without Water: The Scramble to Stop Jordan from Running Dry." *Nature* 573 (7772): 20–23. https://doi.org/10.1038/d41586-019-02600-w.

WHO/UNICEF Joint Monitoring Programme (JMP) for Water Supply, Sanitation and Hygiene. 2021. Progress on Household Drinking Water, Sanitation and Hygiene 2000-2020. Accessed June 3, 2024. https://www.who.int/publications-detail/progress-on-household-drinking-water-sanitation-and-hygiene-2000-2020

5 Analysis of Climatic Extremes in the Gambia Watershed in the Context of Hydroelectric Infrastructure Development

Abdoulaye Faty

5.1 INTRODUCTION

The Gambia River Development Organisation's (OMVG) area of interest covers the three catchments' areas of the Gambia, Kayanga-Géba and Koliba-Corubal rivers (Figure 5.1). Together they represent a surface area of just over 116,000 km², which discharges an average of 20,000 mm³ of water into the Atlantic Ocean each year. The behaviour of the Senegal River watershed is characterised by its hydraulic capacity, which is highly dependent on rainfall, which varies between 1,200 and 1,500 mm per year, and topography (800–1,200 m altitude). By way of illustration, the Corubal sub-basin records an average rainfall of 1,250 mm per year and is largely located in the mountains, which gives it very good hydraulic capacity; the Gambia basin has much lower flows; and the Rio Geba is a lowland river and has even lower flows. Other physical aspects can be summarised as follows:

Abundant rainfall of uncertain climatological character: With the exception of the Nieriko and Sandougou basins, the region receives more than 1,000 mm of rain per year everywhere and up to 2,200 mm at the western foot of the massif in the Tominé basin; these quantities of water fall between April and October, or even November. Since the 1990s, rainfall has returned to its long-term average. The main surface water systems are characterised by an abundance of resources and irregular flows. The hydraulic capacity of the basins is variable but always significant. The regional context shows that the specific flows are much higher in the south.

A lithology that explains abundant runoff: the upper basins located in the Fouta Djalon are essentially made up of rocks that are not very porous, making infiltration low or even non-existent. Conversely, the downstream part of

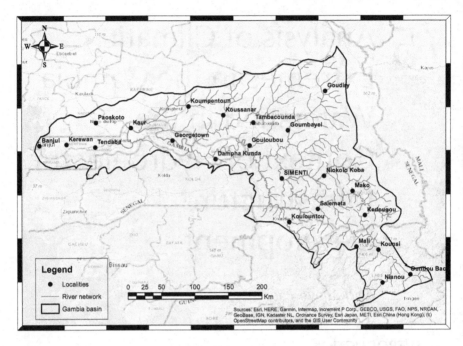

FIGURE 5.1   Map of the Gambia, Kayanga – Géba and Coliba-Korubal watersheds.

Niokolo Koba Park in the Gambia basin, as well as most of the Kayanga/ Geba basin, is underlain by the sedimentary terrain of the Continental Terminal, which has a much higher hydraulic conductivity (and therefore permeability) than the ancient terrain upstream.

As mentioned above, the three basins lie between the Sudanian zone in the north (rainfall between 700 and 1,300 mm per year) and the Guinean zone in the south (annual rainfall above 1,300 mm). This results in a clear differentiation in terms of plant cover; their distribution is therefore primarily zonal but also depends on altitude, exposure, lithology, the presence of watercourses, etc.

It is interesting to note that since the end of the 20th century, there has been a general greening of the Fouta Djallon plateaus and the Sudanian savannahs, which is widespread throughout Sudano-Sahelian West Africa, in connection with the return of rainfall to average levels (Descroix 2018). Moreover, in the Fouta Djalon massif, biomass has been growing steadily since 2000. However, there are pockets of latent deforestation (such as the deforestation halo around the city of Niamey in West Africa). Vegetation cover in West Africa is not systematically decreasing, even though the population continues to grow at a rapid rate (Descroix 2018). Fouta Djallon is one of the best examples of this trend. A negative, Afro-pessimistic belief often associates African deforestation with population growth. This stereotype anticipated catastrophic scenarios based on extensive farming practices, which, in a context of high population growth, would threaten the integrity of Fouta Djallon's ecosystems and thus jeopardise the entire "West African water tower". In fact, local research has opened the debate on the reality and validity of such claims (André and Pestaña 2002). Brandt et al. (2017) have shown localised reductions in woodland areas across West Africa, but that agricultural land in Sahelian zones also promotes woody cover, contradicting simplistic ideas of a strong negative correlation between population density and woody cover.

The watersheds under the jurisdiction of the *"Organisation pour la Mise en Valeur du Fleuve Gambie"* (OMVG) cover a total area of 118,000 km²:

- 78,000 km² for the Gambia River,
- 25,000 km² for the Koliba-Corubal,
- 15,000 km² for the Kayanga-Géba.

OMVG's main mission thus revolves around the rational and harmonious management of the shared resources of the Gambia, Kayanga-Géba and Koliba-Corubal River basins.

In the Gambia River watershed (Figure 5.1), rainwater harvesting has evolved from an informal method (using dishes, buckets, basins, or simply the water accumulated in the foliage of certain plants) to that used in the design of major hydroelectric facilities (turbine dams for energy production). The use of permanent installations with storage systems can be the result of traditional practices or, on the contrary, the result of development actions and/or emergency interventions with the introduction of new, totally imported techniques. This explains why, in the Gambia watershed, rainwater harvesting has moved on from traditional methods to new approaches to sustainable development in order to recover rainwater to fill hydroelectric dams and produce clean hydroelectric power to ensure economic development in the study area.

5.1.1 Literature Review (Background and Rationale)

In addition to studies on the hydroclimatic evolution of basins (Bader et al., 2020; Bodian et al., 2018; Descroix et al., 2020; Lamagat et al., 1990; Sambou et al., 2018b, Faty et al, 2021), a number of studies on the monitoring and assessment of the climatic impact on water resources have been developed as part of impact studies for hydroagricultural and hydroelectric developments.

On the Gambia River, several studies have been carried out to assess the influence of the Sambangalou hydroelectric scheme and strategies to minimise its impact downstream (Bader, 2003). The databases critiqued and reconstructed over the period 1954–2017 and the good chronicles available at Kedougou provide a good understanding of the hydrological regime at Sambangalou. The mean annual modulus is estimated at 101 m³/s in Sambangalou, ranging from 32 m³/s in 1985 to 206 m³/s in 1937.

After the end of the rains, the river continues to receive groundwater inflows for 2 to 3 months, then flows drop rapidly from November onwards. At high water, this dam management will therefore reduce the river's flood level during reservoir filling, and the impact of this dam management is estimated at between 1.17 m at Mako and 2.66 m at Wassadou on the monthly median levels in September. In the Niokolo-Koba park, studies have shown that the river communicates with the park's basins/mares, and the drop in level, estimated at 2.09 m at Simenti, would result in insufficient feeding of the basins/mares, a reduction in dry-season grazing areas for herbivores and, in turn, their predators, and a reduced presence of wetland birds.

At low water, given the residual inflow under natural conditions (virtually nil between mid-February and mid-May), this power generation regime will maintain an average low-water flow of around 60 m3/s (equal to the turbined flow). The corresponding rise in low-water levels would be between 1.16 m at Kédougou and 2.12 m at Wassadou (1.61 m at Simenti in the park) under regulated conditions. The effect on water levels in the estuary will again be slight, given the tidal range, but "the mass of fresh water brought in by the river will considerably slow the rise of the salt front from January onwards". Thus, under regulated conditions, "this front, which on average reaches as far as KP 250 (Kuntaur in Gambia) and even further in dry years, will peak between February and May at KP 155, downstream of Elephant Island, affecting the natural regime of the mangroves.

On the Kayanga/Geba, several studies (Sambou et al, 2018b) provide data and knowledge on small-scale developments in the north of the basin, notably on the Anambé, the Confluent (1984) and Niandouba (1997) dams over the period 1998–2001, on the assumptions used by the RIBASIM software operated by the Société d'exploitation agricole de l'Anambé for planning irrigated agriculture activities, and on salt tongue issues and major bed sedimentation. In terms of modelling, the report "Plan d'action GIRE" (PAGIRE) pour la Kayanga Geba provides an exhaustive inventory of hydro-agricultural resource mobilisation structures and lists 19 possible sites (Sambou et al., 2018b). On Koliba/Corubal, a feasibility study for Saltinho was carried out by the Portuguese company COBA in 1993, but a detailed environmental and social impact study has been commissioned in 2019 (to COBA and Artelia) to update this study.

5.2   RESEARCH METHODOLOGY

5.2.1   Methods

Two approaches were applied to assess drought in the Gambia River catchment: the Standardised Precipitation Index (SPI) and the Climatic Moisture Index (CMI), which provide a quick overview of phenomena as complex as hydro-climatic variability. Both approaches facilitate the comparison of situations at different periods and thus the identification of possible evolutions. The data were processed using Khronostat software. The results were spatialised with the Surfer software.

5.2.1.1   Standardised Precipitation Index

The SPI is based on the following formula:

$$SPI = (Xi - Xm) / Si \qquad (5.1)$$

where:

   Xi is the cumulative rainfall for a given year
   I, Xm and Si are, respectively, the mean and standard deviation of the annual rainfall observed for a given series.

The precipitation index defines the severity of drought in different classes. Negative annual values indicate drought compared to the chosen reference period, and positive values indicate wetness.

5.2.1.2 Climatic Moisture Index (CMI)

For an understanding of water stress in the basin, monthly rainfall evapotranspiration datasets were used to extract information through the CMI. The CMI was calculated as the difference between annual precipitation and potential evapotranspiration (PET) – the potential loss of water vapour from a landscape covered by vegetation. Positive CMI values indicate wet or moist conditions and show that precipitation is sufficient to sustain a closed-canopy forest.

5.2.2 PRESENTATION OF THE STUDY DATA

The question of rainfall trends is of prime importance in the Sahel and particularly in the Gambia watershed, a region where rainfall is at the heart of society's concerns. Deficits affect the availability of water resources and the yields of what is still predominantly rain-fed agriculture. Excessive rainfall can lead to extreme hydrological events, which are detrimental to populations increasingly exposed to the risk of flooding. Rainfall is also the signature of the atmospheric and environmental processes that regulate the West African monsoon, itself a component of the global climate system. Rainfall trends, therefore, lie at the interface between climate variability and its impact on populations. Characterising rainfall trends is essential for understanding water-related risks and anticipating their future in a changing global climate. This scientific contribution presents rainfall data from eight rainfall stations, namely Gaoual, Labé, Mamou, Koundara, Mali, Kédougou, Tambacounda and Vélingara. They have rainfall data series collected over more than 50 years (from 1955 to 2019) and offer good coverage of the Gambia basin. Data were homogenised using the double-cumulation method, and gaps were filled using linear regressions between neighbouring stations. The data set was supplied by the "Organisation pour la Mise en Valeur du fleuve Gambie" (OMVG).

5.3 RESULTS

5.3.1 RAINFALL STUDY

West Africa has experienced three notable rainfall phases since the middle of the 20th century:

From 1950 to 1967, a hyper-humid episode.
From 1968 to 1993 (from 1968 to 1998 in western WA and therefore the area of interest here), a long period of rainfall deficit, very pronounced in the Sahelian zone but noticeable even in the Guinean and coastal regions.
Since 1994 (1999 in the west and therefore in the basins of concern to us here), a period of average rainfall.

Water resources have been severely impacted by the rainfall deficit phase, which has been long and pronounced and has affected huge areas of West Africa. Most major rivers have not yet recovered their pre-drought flows. The return of annual rainfall to its long-term average occurred 5 years later in West Africa (in 1999) than in West Africa as a whole (return in 1994 to the long-term average).

5.3.2 HYDROLOGICAL AND HYDROGEOLOGICAL ANALYSES

A great similarity between the annual flows of the three rivers and the rainfall was observed: a decrease in flows until the 1980s and a clear increase from 1994 onwards. Three key sectors should be noted for their particular hydrological systems:

Two sectors with very high rainfall:
The sector of the three springs, located 15 km north of Labé, constitutes a mole of hard and impermeable rocks. This mole receives abundant rainfall (between 1,500 and 1,600 mm per year) due to its excellent exposure to monsoon flows and has a very low water retention capacity, hence the high flow coefficients observed in the high basins that cut it.
The Tougué (or Mali) massif is one of the best-watered points in the FD; the average annual rainfall in Mali is 1,720 mm.
A sector with a surprising capacity to retain rainwater and release it during the dry season is the Badiar "massif" (or sector of the three borders).

5.3.3 ANALYSIS OF INTER-ANNUAL PRECIPITATION

In most studies, annual rainfall anomalies are estimated by calculating a SPI (Ali and Lebel, 2009). The SPI makes it possible to distinguish between years above and below the climatological average and to derive the main trends in rainfall from inter-annual and decadal variability.

In the early 2000s, the question of whether the drought would continue, or not, was the subject of debate. By analysing the annual rainfall SPI for the Sahel up to 2000, Hote et al. (2002) argue that dry conditions are still predominant at the end of the 1990s (Figure 5.2). Ozer et al (2003) refute these conclusions, claiming, on the contrary, that the drought ended in the mid-1990s. Extending the study period to 2003, Dai et al (2004) and Nicholson (2005) note a trend towards increased rainfall in the Sahel, particularly in the Gambia watershed.

Figure 5.2 shows the evolution of annual rainfall for several stations with long observation series in the three OMVG basins. However, the station with the highest rainfall (Gaoual) had an interrupted series after 1982; the second wettest station, Mali, had a gaping series from 1983 to 1990. It is also the highest rainfall station in Fouta Djalon, the town being located at an altitude of 1,482 m.

For all stations, a surplus period is observed from 1950 to 1967; for the stations in Guinea (Gaoual, Labé and Mali), a first surplus period appears from 1925 to 1936. Then, for all stations, the dry period is observed from 1968 to 1993 (or 1998). Apart from Gaoual (Figure 5.3), where the series ends in 1982, the return of rainfall to its long-term average at the end of the 20th century can be observed.

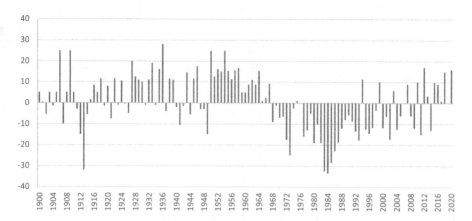

FIGURE 5.2 Changes in annual rainfall since 1951 in Senegambia: standardised precipitation index (SPI), variation in the number of standard deviations from the mean.

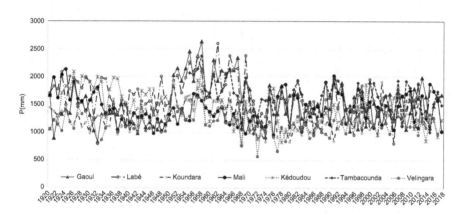

FIGURE 5.3 Changes in annual rainfall since 1920 at selected stations with the longest data series in the OMVG basins.

5.3.4 ANALYSIS OF SPI

The SPI makes it possible to characterise deficit years and surplus years with greater precision and to deduce the main trends over the period in question. From 1969 onwards, deficit years became very recurrent. The successive frequency of the reversal of rainfall trends explains the onset of the drought cycle in the Gambia catchment. Figure 5.4 shows average index values over different periods between the surplus period (1955–1964) and the deficit period (1970–2014). The CMI also shows a predominantly dry situation. Figure 5.4b effectively confirms the drought character of the Gambia River catchment. Most of the CMI values fall within the range of semi-arid conditions ($-0.6 < CMI < 0$). Finally, arid conditions are much less frequent. This is in line with the results. Averaging all stations, the values go from -0.16

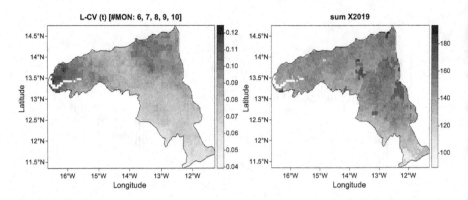

FIGURE 5.4 Variation in precipitation index (a) and drought (b).

to −0.27 over the decades of the period 1955–1969 to values below −0.35 (−0.36 to −0.49) over the decades from 1970 to 2014. Spatially and temporally, the drought is more persistent in the northern area than in the south.

5.4   DISCUSSION

Knowledge of water resources is based on an initial diagnosis of the rainwater network. Given the hazards of climate change (variations in rainfall), populations are resorting to rainwater harvesting for domestic and economic purposes (creating reservoirs to store rainwater for electricity generation).

From one end of the planet to the other, the situation is the same: the growing world population, combined with a steady increase in water abstraction (more than doubling between 1950 and 1990), is putting enormous pressure on drinking water resources. It is estimated that the quantity of renewable fresh water available per capita fell from 17,000 m^3 in 1950 to 7,500 m^3 in 1995 and is set to fall to 5,100 m^3 by 2025. By 2030, demand could even exceed supply (DSDS, 2006). Against this backdrop, rainwater harvesting is increasingly seen as a promising alternative. Japan, one of the first countries to develop installations, has been followed by Canada, India, Brazil and Ethiopia. In Europe, Germany, which has been moving in the same direction for some 20 years, benefits from a standard on the subject and an organised professional sector.

In some developing countries, such as Comoros, the most isolated villages or "campuses" have no access to the drinking water distribution system. Boreholes are far from inhabited areas, or even non-existent. The only available source of water is the river, so the populations of these villages, or "campuses", have developed rainwater harvesting and storage systems to meet household water needs, particularly for drinking. The water is collected from the roofs into containers such as buckets, basins, petrol barrels or tuff tanks. The tuff tank is the most widely used, and it is sometimes shared between homes. It may or may not be fitted with a "home-made" filtration device, which is usually a simple piece of cloth.

According to Pelissier (1958), in the ancient kingdom of Bandial, which today corresponds to the rural community of Enampore (west of Ziguinchor), the architecture of dwellings was organised around a central courtyard where rainwater was collected and stored. Rainwater is channelled into an earthen trough, from which it drains to the outside via a buried drain made of roan (a wood reputed to be rot-proof and very strong). The impluvium huts, a genuine cultural heritage, are currently the subject of restoration projects.

5.5 CONCLUSION

This chapter has updated the climatic impact on water resources over the last 20 years in the context of hydroelectric infrastructure development in the Gambia River watershed. The main point to note is that the deficit in the number of rainy days has persisted over the last decade but has been offset by a greater occurrence of heavy rains associated with the most widespread rainfall systems. This change in rainfall patterns has had a twofold consequence: a return to better annual rainfall, with the greater occurrence of heavy rains more than offsetting the persistent deficit in the number of rainy days, and an unprecedented modulation of the decadal rainfall signal by the intensity of the heaviest rains. Climate change is an established phenomenon, generating uncertainties about the availability and sustainability of water resources in the Gambia River watershed. In this context, the use of rainwater is becoming increasingly important, as highlighted at the 9th World Water Forum in Dakar (Senegal). To anticipate the effects of climate change and try to reduce vulnerability to its negative impacts (droughts, etc.), rainwater harvesting can be one of several possible solutions. Taking into account the different uses and needs for water, rainwater harvesting should be considered as part of a broader strategy of integrated water resource management: how to optimise the different water resources according to their availability and different uses.

FURTHER READING

Available data are listed in several studies on the Gambia zone (Bader, 2003; Bader et al., 2020; Bodian et al., 2018; Descroix et al., 2020; Lamagat et al., 1990); on the Geba (Sambou et al., 2018b); and on the Koliba (Albergel and Pépin, 1991; Sambou, 2019). Many data are available on Aquacoope, a data platform born from the data digitisation and visualisation initiative of the Geneva Water Hub and UNCDF.

REFERENCES

V. André, G. Pestaña, 2002. Les visages du Fouta-Djalon. *Les Cah. d'Outre-Mer, 217.* Disponible en ligne: https://com.revues.org/index1038.html

J.C. Bader, H. Dacosta, J.C. Pouget, 2020. Water (Switzerland) 12, 2520.

A. Bodian, A. Dezetter, L. Diop, A. Deme, K. Djaman, A. Diop, 2018. Future climate change impacts on streamflows of two main West Africa river basins: Senegal and Gambia. *Hydrology* 5, 21.

M. Brandt, K. Rasmussen, J. Peñuelas, F. Tian, G. Schurgers, A. Verger, O. Mertz, J.R.B. Palmer, R. Fensholt, 2017. La croissance de la population humaine compense l'augmentation de la végétation ligneuse due au climat en Afrique subsaharienne. *Nature Ecology & Evolution, 1*, 0081.

L. Descroix, 2018. *Processus et Enjeux d'eau en Afrique de l'Ouest Sahélo-Soudanienne;* Editions des Archives Contemporaines: Paris, France, 320p, ISBN 9782813003140.

L. Descroix, B. Faty, S P Manga, A. B. Diedhiou, L.A. Lambert, S. Soumaré, J. Andrieu, A. Ogilvie, A. Fall, G. Mahé, F. B. S. Diallo, A. Diallo, K. Diallo, J. Albergel, B. A. Tanimoun, I. Amadou, J.-C. Bader, A. Barry, A. Bodian, Y. Boulvert, N. Braquet, J. L. Couture, H. Dacosta, G. Dejacquelot, M. Diakité, K. Diallo, E. Gallese, L. Ferry, L. Konaté, B. N. Nnomo, J. C. Olivry, D. Orange, Y. Sakho, S. Sambou, J.P. Vandervaere, 2020. Are the Fouta Djallon Highlands still the water tower of West Africa? *Water* 12, 2968; doi:10.3390/w12112968.

Direction de la santé et du développement social de la Guyane et NBC (Octobre 2006). Etude sur la récolte d'eau de pluie pour l'usage alimentaire dans les sites isolés de la Guyane.

A. Faty, F. Kouame, A. N. Fall, A. Kane, 2021. Land use dynamics in the context of variations in hydrological regimes in the upper Senegal River basin. International Journal of Hydrology. 2019;3(3):185–192.

J.P. Lamagat, J. Albergel, J.M. Bouchez, et L. Descroix, 1990. Monographie Hydrologique du fleuve Gambie. Orstom-OMVG, Monographie Orstom, Orstom-Dakar, 246 p.

D. Orange., 1990. Hydroclimatologie du Fouta Djallon et dynamique actuelle d'un vieux paysage latéritique. PhD thesis, Université Louis Pasteur de Strasbourg, France, 232 p.

P. Pelissier, 1958. Les Diolas: étude sur l'habitat des riziculteurs de Basse-casamance, les Cahiers d'Outre Mer N° 44, pp 334–388

S. Sambou, 2019. Contribution à la connaissance des ressources en eau du bassin versant du fleuve Kayanga/Gêba (Guinée, Sénégal, Guinée-Bissau). Thèse de doctorat, Université Cheikh Anta Diop, 314 p.

S. Sambou, H. Dacosta, J.-E. Paturel, 2018b. Variabilité spatio-temporelle des pluies de 1932 à 2014 dans le bassin versant du fleuve Kayanga/Gêba (République de Guinée, Sénégal, Guinée-Bissau). Physio-Géo. 2018;12:61-78. Available from: http://journals.openedition.org/physio-geo/5798

Case Study 1

Demonstrating Decentralised Circular Rainwater Management Solutions to Combat Water Scarcity in the Mediterranean

The HYDROUSA Mykonos Demo

K. Monokrousou, C. Makropoulos,
A. Eleftheriou, I. Vasilakos, M. Styllas,
K. Dimitriadis, and S. Malamis

SUMMARY

The HYDROUSA Mykonos case study is a decentralised, flexible and autonomous rainwater management system that includes aquifer storage and recovery schemes. This prototype has been developed and implemented on the highly touristic Mediterranean island of Mykonos (Cyclades, Greece). The approach makes optimal use of low-cost rainwater/stormwater through a residential rainwater collection system and nature-based storage afforded by the subsurface, with a positive impact on the environment. The overall solution comprises several components, mainly existing infrastructure, in order to maximise the utilisation of natural water resources, enable the buffering effect and extend water availability well into the dry period for domestic and agricultural water reuse at the point of demand. The general principle is to make the most of the infrastructure of the system – tanks, terraces and bioswale – to collect and store rainwater for domestic use, irrigation and aquifer recharge.

DOI: 10.1201/9781032638102-7

This novel system has been developed to test how to address water scarcity issues, which tend to be more significant, especially in regions suffering from a lack of water availability and irregular distribution of their water reserves. Furthermore, the increasing tourist activity during the summer puts additional stress on water supplies, both in terms of quantity and quality. For this purpose, we have implemented such a solution on a Mediterranean island that faces great challenges in terms of water availability in order to demonstrate a solution 'bundle' that is scalable and replicable and can form part of flexible and resilient regional climate change adaptation pathways, especially suitable for the European South.

PROBLEM STATEMENT

As water scarcity becomes more prevalent in semi-arid regions, new circular water management systems are needed to ensure sustainable environmental and economic development in these regions. The solution that has been implemented to address such problems is a novel water management system. It practically collects rainwater from rainfall events, stores it temporarily in tanks and in the subsurface and recovers it for reuse when it is needed. As such, rainwater can indeed be stored and gradually recovered from the subsurface through an optimised, autonomous, flexible and decentralised water recovery system in real-world environments. This system can potentially be upscaled and replicated in similar water-scarce areas, providing a 'climate-proofed' solution.

BENEFICIARIES

There are a number of beneficiaries in this demo case. Farmers, farmers' associations and cooperatives can benefit from water savings, cost reductions and increases in agricultural production. Municipalities and water utilities can use such solutions as they allow for better and more efficient water services. These configurations can be implemented on larger scales in decentralised or remote areas to address the costly matter of water availability through transferring water (either through a water supply network or through the construction of major dams, etc.). Especially in regions like the Mediterranean Sea, with numerous islands (about 6,000–10,000 islands and islets) and remote small settlements where there are no water supply networks available, such solutions can promote sustainability. The public, NGOs and schools can be inspired by such interventions as they can convey the message of the circular economy (CE), sustainability and climate change adaptation solutions. These solutions can also be beneficial for hotspots with refugees and immigrants that are constructed in decentralised areas with limited water resources. Lastly, these solutions can provide valuable insights to water regulators and policymakers. The results and outcomes of these projects can be utilised to inform the decision-making process, leading to legislative adjustments that foster sustainable water use and CE.

ACTIONS TAKEN

The HYDROUSA Mykonos case study consists of two separate subsystems. The first, referred to as a domestic rainwater collection system, collects rainwater from concrete roofs with a total area of $438\,m^2$ through gutters, which is led into a storage tank with a volume of $70\,m^3$ to meet the non-potable needs of households. In the second subsystem, the aquifer storage and recovery system, rainwater is collected from the surface runoffs inside the site (approximately $350\,m^2$) and transported by gravity to a storage tank with a capacity of $40\,m^3$. The collected water is used for aquifer recharge and irrigation purposes on the adjacent land (0.2 hectares of lavender). Additionally, in the second subsystem, rainwater is collected through an open-channel linear drainage system (a bioswale system), stored in an open tank with a capacity of $20\,m^3$, and in the subsurface "basin" in the location of the artificial recharge (AR) well. The bioswale system not only collects and partially treats storm water but also prevents the field from flooding. When water is needed for irrigation, it is recovered primarily from an AR well and secondarily from the storage tank or the open tank. Subsequently, when water is recovered in the AR well (some days later), this is sent back to these tanks so that they are always full. All systems are fully automated with a Programmable Logic Controller (PLC) system installed in the site, and analysis tests are also conducted periodically to assess water quality.

RESULTS

To provide a thorough understanding of the hydraulic behaviour of the aquifer acting as the buffer for water storage and to evaluate the aquifer water retention capacity, data on rainfall events have been taken from onsite sensors (quantity and quality), complemented by periodic manual measurements as well as from two AR experiments conducted at different times of the year (June 2019 and November 2021). The main results are summarised below.

- Rainfall events drive the recharge of the site aquifer through two main mechanisms, with different response times to the events that act synergistically and depending mainly on the precedent and antecedent meteorological conditions.
- The total water volume that can be stored in the subsurface of the AR area is about $1{,}050\,m^3$.
- Stored water can be retrieved at depths of $4\,m$, where a new well (AR well) has been constructed.
- There is a $40\,mm$ threshold of rainfall for the gradual natural recharge of the aquifer, giving a lagged rise to the groundwater level of the AR well.
- For about $1\,m^3$ of water recovery for irrigation, a water level drop of $1.8\,m$ in the AR well is recorded. Then a rise of about $1\,m$ is gradually achieved in about seven days, suggesting a significant recharge potential of the aquifer.

- The water quality is satisfactory in terms of electrical conductivity and sodium hazard; no significant variation is noticed.
- A proof-of-concept has been established that rainwater can be stored both on the surface and underground and gradually recovered when it is needed.

SUSTAINABLE DEVELOPMENT GOALS AND IMPACT

The project addressed the following Sustainable Development Goals: SDG 6: clean water and sanitation; SDG 11: sustainable cities and communities; SDG 12: responsible consumption and production and SDG 13: climate action.

The HYDROUSA Mykonos decentralised rainwater management system is an innovative effort to mitigate future aridity trends in the Mediterranean. The design, construction and operation of this rainwater management system have resulted in

- testing a transformative adaptation water management solution in real world water scarce environments within the concept of CE.
- increasing community resilience in remote, decentralised settlements.
- adapting to the projected climate change and carrying out a sustainable transition.

The combination of nature-based subsurface water storage and rainwater harvesting from existing infrastructures makes the applicability of the prototype widespread, not only at household level but even in larger properties such as hotels and buildings located off the bounds of towns. Larger properties have larger roofs and paved surface areas, thus higher storage potential in existing water tanks or in the more extended subsurface zone. Such a system may not be able to accommodate all the needs of small and large tourist infrastructures. However, an upscaling experimental effort will further optimise the system and provide additional benefits to the local communities.

The main impact of this work is the fact that this configuration has been implemented and tested in a real-world setting, and the results aim to serve as a step towards more flexible, expandable, scalable and replicable CE solutions.

NEXT STEPS

The next steps to be implemented in the case study include:

- Continuous monitoring of the site over one more complete hydrological year will result in a statistical model that will feed the site subsystem decision process for optimising water storage and recovery in the aquifer system.
- Modelling activities for the simulation of the aquifer systems to support the optimisation of the subsystems and calibration under steady-state conditions based on newly acquired field data.
- Optimising the setup of the configuration as well as the retention potential of the subsurface "basin".

ADVICE FOR OTHERS LOOKING TO DO SOMETHING SIMILAR

One of the main considerations in such projects is the local geology. The underlying geology of the test site on Mykonos Island is composed of crystalline granitic bedrock. Non-destructive geophysics (Electrical Resistivity Tomography (ERT) and Ground Penetrating Radar (GPR)) confirm the existence of a 4-m-thick weathered zone, which serves as the buffer zone where water can be stored. Crystalline bedrock subsurface topography can be easily delineated by ERT and GPR, but groundtruthing through the excavation of a soil profile or an exploratory well, which can later be used as the AR well, is necessary. In the case of carbonate bedrock, it is likely that additional geological and geophysical surveying will be needed to define layers of high hydraulic conductivity and the geometry of the subsurface aquifer. Since most households in the Mediterranean islands have existing wells, all existing historical information on the recharging capacity must be collected and lead to a thorough understanding of the recharging potential of the subsurface aquifer. After the geological assessment of the underground conditions, the stakeholder needs to obtain a thorough understanding of the hydraulic behaviour of the aquifer selected to act as the buffer for water storage and to evaluate the aquifer water retention capacity through multiple recharge tests both in the summer and winter seasons. It is also crucial to estimate the rainwater harvesting capacity from roofs, surface runoff, etc.

BIBLIOGRAPHY

Cramer, W., Guiot, J., Fader, M. Garrabou, J., Gattuso, J-P., Iglesias, A., Lange, M.A., Lionello, P., Llasat, C.M., Paz, S., Peñuelas, J., Snoussi, M., Toreti, A., Tsimplis, M.N., Xoplaki, E. 2018. Climate change and interconnected risks to sustainable development in the Mediterranean. *Nature Climate Change*, 8: 972–980 https://doi.org/10.1038/s41558-018-0299-2

Giorgi, F., Lionello, P. 2008. Climate change projections for the Mediterranean region. *Global and Planetary Change*, 63: 2–3, 90–104 https://doi.org/10.1016/j.gloplacha.2007.09.005

Makropoulos, C., Nikolopoulos, D., Palmen, L., Kools, S., Segrave, A., Vries, D., Koop, S., Van Alphen, H.J., Vonk, E., Van Thienen, P., Rozos, E. 2018. A resilience assessment method for urban water systems. *Urban Water Journal*, 15: 4, 316–328

Tahir, S., Steichan, T., Shouler, M. 2019. *Water and circular economy: a white paper.* Ellen MacArthur Foundation, ARUP, Antea Group

Ulbrich, U., Lionello, P., Belušić, D., Jacobeit, J., Knippertz, P., Kuglitsch, F.G., Leckebusch, G.C., Luterbacher, J., Maugeri, M., Maheras, P., Nissen, K.M., Pavan, V., Pinto, J.G., Saaroni, H., Seubert, S., Toreti, A., Xoplaki, E., Ziv, B. 2012. Climate of the Mediterranean: Synoptic Patterns, Temperature, Precipitation, Winds, and their Extremes. *The Climate of the Mediterranean Region.* Elsevier, 301–346. https://doi.org/10.1016/B978-0-12-416042-2.00005-7

Case Study 2

Rainwater Harvesting as an Affordable Drinking Water Solution for Emerging Chemical Contaminants

A Case Study of Guanajuato, Mexico

Dylan Terrell

SUMMARY

The Alto Río Laja Watershed stretches across seven municipalities in northern Guanajuato State in central Mexico and is a microcosm example that illustrates many of the extremely complex water quality and scarcity challenges facing Mexico, as well as many other parts of the world, today. Almost all of the water consumed in this region comes from a large underground reservoir known as the Alto Río Laja Aquifer, which serves more than 700,000 residents across several thousand distinct communities, rural and urban alike. The aquifer is declining at an alarming rate, from 1 to 3 m per year (Hoogesteger & Wester, 2017). As a result, community wells are drying up and, in many cases, literally collapsing in on themselves.

Further complicating the issue, the water that does remain is often contaminated with arsenic and fluoride – up to 23 times the World Health Organization's recommendation for arsenic and more than 18 times the Mexican limit for fluoride (Caminos de Agua, 2023). These extremely hard-to-remove contaminants are closely linked to dental fluorosis, crippling skeletal fluorosis, chronic kidney disease, cognitive development and learning disabilities in children, skin disease, and various cancers (Farías et al., 2021; WHO, 2022).

Since 2012, the NGO Caminos de Agua (Caminos) has been working in partnership with dozens of rural communities, grassroots organisations and community

DOI: 10.1201/9781032638102-8

leaders, academics, government, and other institutions and experts to monitor regional water quality, provide educational programming, and innovate and implement low-cost water solutions to these complex water challenges. The organisation's rainwater harvesting programmes have been paramount to their impact and help to improve community health, reduce environmental stress on over-extracted aquifers, and give users control and consistency over their water supply.

PROBLEM STATEMENT

Globally, emerging chemical contaminants, both synthetic and naturally occurring – like arsenic, fluoride, lead, and pesticides – are finding their way into local water supplies and disproportionately impacting low-income communities. In Mexico alone, an estimated 21 million people are exposed to excessive levels of arsenic and/or fluoride (Del Razo et al, 2021), with few appropriate solutions available. These chemicals are often hard and expensive to identify, and even more so, to remove. Rainwater harvesting, on the other hand, is a water source that is inherently free of hard-to-remove chemical contaminants like arsenic and fluoride. When combined with a basic water filter to remove biological pathogens, rainwater becomes an affordable source of clean drinking water.

In total, as of 2023, Caminos de Agua's projects have directly impacted roughly 45,000 direct beneficiaries across Mexico and other parts of Latin America through their Aguadapt water filters, community-scale water treatment plants, and composting toilets, with an estimated 7,500 people gaining clean drinking water access through their rainwater harvesting projects in more than 120 mostly rural, low-income communities, including several indigenous communities, across northern Guanajuato State in central Mexico.

ACTIONS TAKEN

Caminos has worked with over 120 different communities, alongside numerous grassroots and community partners, to implement rainwater harvesting systems. Each community is unique, and actions vary from project to project. That said, rainwater projects tend to begin with weeks of community organisation and a community education programme that allows the local residents to understand the issues with the water supply in the region, and specifically their community and their own wells (Caminos conducts comprehensive water testing on community wells). Further, the education program provides all the training and information that the communities will need to be able to construct and maintain their own rainwater harvesting systems.

Once construction is ready to begin, Caminos first provides a 3–7-day technical capacity training (depending on the model) – usually in a shared space like a school, church, or community centre – where community members build a full-scale rainwater harvesting system. From there, the community organises itself, often in small workgroups or as one large group, to build the rest of the systems. Caminos provides oversight and technical assistance, but by having the residents themselves build the systems, a sense of ownership is often engendered, which greatly improves the likelihood that the systems will be used and maintained over time. The systems include gutters, tubing, and downspouts to capture the rainwater; a "first-flush" system that separates out the first rains, leaf filters, and finally, a large-capacity (usually 5,000–30,000-L depending on the project) cistern for long-term water storage. Often, the systems are constructed with 12,000-L ferrocement cisterns. Each household also receives one of Caminos' certified Aguadapt water filters to remove any biological pathogens and to assure the rainwater is safe for human consumption. Each rainwater harvesting system is designed to provide between 1 and 3 families, or an entire school, with decades of clean water for drinking and cooking.

Caminos also implements an extensive Monitoring and Evaluation (M&E) Program, based on internally developed software and surveys to assure the long-term success and sustainability of their projects. After initial baseline surveys with all the participating families, we return to do follow-up surveys as well as technical evaluations of the systems. The M&E Program measures dozens of indicators across eight different objectives and helps us to continuously improve our educational programmes, community organisational strategies, and implementation methods.

RESULTS

Rainwater harvesting systems – like any technology – are useful tools but not complete solutions in and of themselves. Success is dependent on community participation and ownership. For Caminos' rainwater projects, community members make all decisions regarding organisation, beneficiaries, and locations. Caminos utilises a unique community organisation methodology, Narrative Practices, to help engage community members to tell their own stories and take ownership over the process.

Paramount to success lies in the close partnership with grassroots and community organisations, specifically Pozo Ademado Community Services (SECOPA), The San Cayetano Community Center, and United Communities for Life and Water (CUVAPAS), which represent dozens of rural communities throughout the watershed.

Caminos has developed a rigorous M&E program to measure the impact of its programmes, which has shown concrete results. For example, the prevalence of water-derived health conditions, such as dental fluorosis in children, has been measurably reduced from generation to generation. Children born after the installation of rainwater systems no longer suffer from severe stomach pain and dental fluorosis that were prevalent in the community before the interventions. We have also reduced arsenic and fluoride-contaminated water consumption by 100% in exposed communities and increased overall water access by 26%. In another example, 92% of households had lowered their annual spending on potable water from 22% of their income on average to less than 2%. Promoting access to clean water not only improves health outcomes and reduces the load on the health system, but also develops social cohesion by reducing conflict due to water issues.

SUSTAINABLE DEVELOPMENT GOALS AND IMPACT

The projects implemented impacted most on the following SDGs: SDG 6: Clean water and sanitation and SDG 3: Good health and well-being. To date, Caminos de Agua has partnered with government agencies, other NGOs, and most importantly, numerous grassroots and community organisations to work in more than 120, mostly rural, communities in northern Guanajuato to install over 1,140 large-scale rainwater harvesting systems in community homes and schools that have a combined capacity to create more than 450 million litres of clean drinking water over their lifetime.

NEXT STEPS

Rainwater harvesting remains one of the most actionable technologies Caminos de Agua will continue to implement to combat increasing levels of arsenic and fluoride in Mexico. Between 2020 and 2021, Caminos was able to triple the number of rainwater harvesting systems they were able to install, and they doubled that number again in 2022, becoming increasingly efficient and effective in their programming. Caminos will continue to partner with the government, other NGOs, and, of course, grassroots organisations to continue to scale rainwater capacity moving forward.

That said, there are tens of millions of people impacted by arsenic and fluoride in their groundwater in Mexico alone. These contaminants are often found in arid/semi-arid regions, meaning that rainwater systems need to be sized to store enough water to endure the extensive dry seasons, lasting eight months or more every year. The need is simply too large to meet through rainwater alone. Recognising the need to address these contaminants in ways that serve more people faster, Caminos has also been focusing on scaling their innovative Groundwater Treatment System (GTS) to effectively remove arsenic and fluoride from contaminated groundwater at the community scale. At the same initial cost as one rainwater harvesting system, which serves one family, the first GTS currently in operation is serving 40 families, and by the end of 2023, Caminos began installing a GTS for a community of 270 families. Caminos plans to create a model with GTS that can be replicated by others facing similar water quality challenges around the globe.

ADVICE FOR OTHERS LOOKING TO DO SOMETHING SIMILAR

Do not enter communities with a solution in hand. Work directly with community members and, when possible, local grassroots groups and organisations to fully understand the local reality, culture, and context before making any plans. Partner with these groups to understand all of the potential solutions and come to the choice of moving forward with rainwater harvesting organically. In the end, the decisions should ultimately be made by those who will be using and owning the technology. Also, with any technology, and especially with rainwater harvesting used for drinking water, if you are not confident enough to use it yourself and with your own family, you probably should not be promoting it as a solution to others.

REFERENCES

Caminos de Agua. 2023. "Water Quality Monitoring." Caminos de Agua. https://www. caminosdeagua.org/en/water-quality-monitoring.

Farías, Paulina, Jesús Alejandro Estevez-García, Erika Noelia Onofre-Pardo, María Luisa Pérez-Humara, Elodia Rojas-Lima, Urinda Álamo-Hernández, and Diana Olivia Rocha-Amador. 2021. "Fluoride Exposure through Different Drinking Water Sources in a Contaminated Basin in Guanajuato, Mexico: A Deterministic Human Health Risk Assessment." *International Journal of Environmental Research and Public Health* 18 (21): 11490. https://doi.org/10.3390/ijerph182111490.

Hoogesteger, Jaime, and Philippus Wester. 2017. "Regulating Groundwater Use: The Challenges of Policy Implementation in Guanajuato, Central Mexico." *Environmental Science & Policy* 77 (November): 107–113. https://doi.org/10.1016/j.envsci.2017.08.002.

Del Razo, M., Ledón, J. M., and Velasco, M, N. 2021. Arsénico y fluoruro en agua: riesgos y perspectivas desde la sociedad civil y la academia en México. 1ª Ed.- Ciudad de México, UNAM-Instituto de Geofísica. IGEF_derecho_humano_al_agua_2021_1ed.pdf (unam. mx)

World Health Organization. 2022. "Arsenic." December 7, 2022. https://www.who.int/ news-room/fact-sheets/detail/arsenic#:~:text=Long%2Dterm%20exposure%20to%20 arsenic.

Part 2

Rainwater Harvesting in Practice

6 Scaling-up Rainwater Harvesting in Mexico City
A Socio-Environmental Review

Jorge A. Ortiz-Moreno, Beth Tellman,
Emilio Rodríguez-Izquierdo Michelle Morelos,
and Laura Rodríguez-Bustos

6.1 INTRODUCTION

Water scarcity affects more than 40% of the global population, mainly due to physical shortages, the failure of institutions to ensure a regular supply or a lack of infrastructure (United Nations, 2022). On a global scale, 2.2 billion people still lack safely managed water, including 785 million without access to essential drinking water (Ibid.). Insufficient water, among other things, can limit access to basic hygiene and healthcare facilities or make them more expensive. The water scarcity problem is expected to increase due to climate change and is more common and complex in poor and middle-income countries. In urban settings, water scarcity is often framed as the result of a larger demand that outstrips the available supply. However, the root causes of scarcity often reflect exclusionary processes and unequal power relations shaping access to and control over water resources (Mehta, 2013. Addressing water security is critical for every country, and eradicating water scarcity requires a redesign of policies. As urbanisation exacerbates the social inequalities expressed through geographical differences in access to safe water (Li et al., 2020; McDonald et al., 2011), these inequalities are part of a wider discussion about the equality and sustainability of water supply in cities.

During the past decades, technological innovation has played a central role in urban system designs that ensure human well-being and the sustainable use of water to maintain planetary cycles (Bashar et al., 2018; Mihelcic et al., 2007). In this context, rainwater harvesting (RWH) has been suggested as a practical alternative with the potential to reduce the amount of time and money a family spends on water and improve their access to a safe water source (Karim et al., 2015; Lee et al., 2016). In the process of building a water-saving society (Zheng et al., 2010), RWH systems could be considered a sustainable technological solution that directly addresses the

DOI: 10.1201/9781032638102-10

problems of scarcity. However, one of the relevant methodological challenges is the inclusion of a geographical approach that incorporates spatial differences in hydrological processes, such as rainfall, and socioeconomic factors to adapt the RWH systems to the preferences and possibilities of the beneficiaries (García Soler et al., 2018). Bakir and Xingnan (2008) recognise that spatial and temporal differences in rainfall in cities allow the design and performance of RWH systems to be anticipated, managed, and modified. Therefore, RWH can be more effective in areas with sufficient rainfall levels but which experience access problems and limited availability due to increased water demand and inefficient provision systems (Karim et al., 2015).

The social dimensions and urban spatial configurations are critical to water consumption (Petit-Boix et al., 2018). Likewise, the institutional framework related to political programs and water governance determines the success of public investments and user acceptance (García Soler et al., 2018). In many high-income countries such as the United States, Germany, Spain, Australia, Singapore, France, and Japan, RWH has been legislated at different government levels, even when the water is commonly used for purposes other than human use (Dumit Gómez & Teixeira, 2017). In less developed countries, on the other hand, RWH institutionalisation is scattered as the implementation of this alternative depends mostly on private initiatives, sometimes being the only resource supply infrastructure, and the destination of the water can be for drinking (Haque et al., 2016). However, in countries like Malaysia, India, Brazil, and Mexico, there have been recent efforts led by governmental institutions to support RWH as an alternative water source. Since the 2030 Sustainable Development Agenda points to ensuring the availability and sustainable management of water and sanitation as urgent, the challenges to achieving this depend on the level of commitment from political and civil society, financial incentives, sustainable building codes, and a firm stance of public policy in urban planning (Sivagurunathan et al., 2022; UN, 2022).

In Latin America, there is an awareness of RWH, but it is less trendy, and its acceptance varies depending on the country or region. The most important initiative has been the One Million Rural Cisterns programme (P1MC, Portuguese acronym), through which 1.2 million RWH systems for human consumption were installed in the northeast semi-arid region of Brazil between 2003 and 2016 (Pereira Lindoso et al., 2018). P1MC was environmentally and socially successful in addressing drought problems due to its construction as a local governance initiative between institutions and the population (Ibid.). In Mexico, where only 43% of the population has access to safely managed water services (WHO/UNICEF, 2022), RWH is now growing rapidly as many governments of different levels and different kinds of private institutions have been promoting and disseminating this technology to address water scarcity and inequality. Particularly in the capital of the country, a massive RWH social programme named *Cosecha de Lluvia* (Rainwater Harvest) has been implemented by the city government since 2019. This initiative, coming after more than a decade of RWH institutionalisation efforts, has disseminated over 60,000 RWH systems in nine municipalities in Mexico City (SEDEMA, 2023). However, it is facing challenges in reaching the most vulnerable populations.

This chapter reviews the case of Mexico City from a socio-environmental perspective, discussing the water scarcity problem in the context of climate change, the process of institutionalisation of RWH in the city, the implementation of *Cosecha de Lluvia* from 2019 to 2022, and the complexities of the scaling-up of this technological solution.

6.2   WATER SCARCITY IN THE CONTEXT OF CLIMATE CHANGE

To address water scarcity in Mexico City, it is necessary to consider the current underlying conditions of the phenomenon and its future perspectives, which are expected to worsen due to the consequences of climate change. So far, climate change has modified the temperature and precipitation patterns affecting the Lerma and Cutzamala basins, two of the main sources of water that serve the city. In the last century, the Mexican capital has experienced an increase in mean temperature of 1.6°C, mainly due to the heat-island effect and land use changes, all of which affected the hydrological cycle (Romero-Lankao, 2010). Despite the uncertainty associated with the application of different global climate models, it is projected that precipitation in the Basin of Mexico, where the city is located, will increase in the rainy season (summer) and decrease in the dry season, with a higher frequency and duration of droughts (Martinez et al., 2015). Rainfall projections under climate change scenarios for central Mexico anticipate an expected decrease in annual precipitation for the watersheds upon which the Cutzamala-Lerma system depends, leading to increased drought probabilities (SEDEMA, 2021a).

Larger-scale drought events in the basin of Mexico can compromise up to 28% of the water supply to the city's population. For instance, a drought in 2008 that affected the Cutzamala Basin led to water shortages in 11 boroughs of Mexico City (Martinez et al., 2015). Also, the occurrence of drought months between 2003 and 2008 has increased in Mexico City, going from one month of drought in 2003 to 11 months in 2008 (Escolero et al., 2016). Long-term changes in temperature and precipitation will result in a significant decrease in the availability of freshwater. As a result of climate change, these water shortages will likely affect water access for the most vulnerable people, who have already experienced water precarity for decades. A glimpse of the future was experienced when the most recent drought hit central Mexico in 2021, causing severe consequences for agricultural production, biodiversity and water supply (Arenas Ortiz, 2021).

Traditional water provisioning systems for Mexico City are not able to adequately provide for the city's population under current and future climate change conditions. Despite enormous public investment to build and maintain the infrastructure to import water to the city, 32% of the population (i.e., around 2,800,000 inhabitants) do not have enough water supply to meet their basic needs, and 18% of the city's population is already experiencing water shortages (Torres Bernardino, 2017). Households that depend on water from the Cutzamala-Lerma system are in the west and south of Mexico City, including the boroughs of Magdalena Contreras, Tlalpan, and Azcapotzalco. In these boroughs, vulnerable households that are most exposed to climate-induced drought conditions and the reduction of water supply are also socially marginalised (i.e., low education levels, inadequate living conditions and a lack of health security). These households bear higher costs for water services and are more sensitive to the effects of water insecurity. Women-headed households are particularly vulnerable, as women are usually responsible for allocating water for different domestic uses, recycling water from one use to another, and bearing the costs of anticipating water deliveries, usually at the expense of working days when they are also employed (Eakin et al., 2016).

Furthermore, limited access to locally relevant information is a barrier to implementing adaptation options (Allen et al., 2018). While Mexico City's inhabitants may know about the current water risks that they mitigate, a lack of awareness of future risks due to climate change can prevent citizens from implementing adaptation strategies to address their livelihood vulnerabilities. Thus, access to information on the future effects of climate hazards can change risk attitudes in vulnerable populations by increasing the understanding of the probability and severity of climate risks (Mortreux & Barnett, 2017). Although there are multiple pathways to address the potentially growing water scarcity, most are expensive, politically uncertain and will take time to implement, with no guarantee that they will reach the most vulnerable communities. Given the scale of the issue, it is likely that multiple technologies will be needed to address growing water scarcity. In this light, RWH systems offer several adaptive benefits, taking advantage of natural precipitation, local storage, and flow pathways, and, importantly, can be developed and operated at relatively low cost and can be installed and scaled quickly.

6.3   THE INSTITUTIONALISATION OF RWH IN MEXICO CITY

Environmentalists have adopted domestic RWH since the 1970s as part of the Alternative Technology Movement (Smith, 2005). However, in Mexico City, the urgency of disseminating this practice and technology began resonating in academic and policy forums by the turn of the 21st century as the notion of an impending water crisis putting the city's future at risk became widely acknowledged. Not necessarily challenging centralised technocratic infrastructures as some environmentalists did beforehand, capturing rainwater became increasingly suggested as an alternative to help the city reduce its chronic subsidence originating from the longstanding overexploitation of groundwater sources (Barkin, 2004; Legorreta, 2006). This concern was behind a reform of the Water Law of the Federal District (now Mexico City State) in 2008 that established the incorporation of RWH facilities in new buildings, except for single-family homes. Coincidentally, one year later, different pioneer companies providing RWH solutions for residential, commercial, and industrial purposes were founded. Since then, emerging companies, local government authorities, academic institutions and the media have contributed to the growth of this niche.

So far, various studies have demonstrated the economic and social benefits of scaling up RWH across Mexico City. Concha Larrauri et al. (2019), for example, did a net present value analysis (considering costs of installation, maintenance, and current water rates) of installing RWH systems for domestic use in households, restaurants, and large wholesalers and retailers. They found that the largest cost savings were by wholesalers and retailers (roofs over 2,000 m^2), who could save millions of pesos and cover 7%–15% of their water demand by using rainwater. However, they also found that most domestic households in the city would not save money by installing these systems, except for informal settlements and communities not connected to the grid in the south of the city. These water users, who need to pay for trucked water,[1] could save significant amounts of money by adopting an RWH system. Likewise, a recent study by Wunderlich et al. (2021) found that households with unreliable piped water (~4 hours of supply per week or less) saved money (over 1,000 USD per year) by investing in RWH systems instead of paying for water deliveries. As RWH knowledge is generated in the city, it has become evident that this solution could effectively contribute to addressing the water scarcity experienced by vulnerable populations. Insights that have been useful to inform policymaking.

Institutionalising RWH as a decentralised practice for water supply in Mexico City has followed a bottom-up trajectory, from small borough projects to the multi-year city policy currently under operation. A process that has been strongly influenced by the work of *Isla Urbana*, a social venture founded in 2009 that developed an RWH system model specifically designed for the conditions of a typical single-story peri-urban home. This social enterprise, which has installed more than 25,000 RWH in homes and schools in Mexico City and beyond (Isla Urbana, 2022), soon became the most relevant actor and one of the main local suH knowledge and facility suppliers. Subsidising RWH systems has become a practical solution to partially alleviate the water demand of populations living under water scarcity and social vulnerability conditions, a slightly different motivation than the global environmental issues concerning the early supporters of this practice. In 2010, Tlalpan was the first Mexican borough to invest public funds in a programme that disseminated RWH household facilities. A pioneering initiative that was later replicated and improved by subsequent administrations of the same borough and neighbouring ones, such as Xochimilco, in the city's southern periphery.

As interest in this solution grew in the public discourse, in 2016, the city government finally launched a series of programmes that funded research, equipment and RWH installations in schools, housing units, and public buildings. The city water authority, *Sistema de Aguas de la Ciudad de México* (SACMEX), despite initial reluctance, also incorporated for the first time an administrative unit and a support fund exclusively focused on RWH. When the 2018 mayoral election approached, RWH was already on the public agenda. Four of the seven candidates supported RWH practices during an electoral debate where sustainability and urban issues were addressed (Isla Urbana, 2018). Consequently, the elected mayor, Claudia Sheinbaum, decided to launch *Cosecha de Lluvia*, a massive RWH programme aimed at installing 100,000 household systems in water-stressed and socially marginalised neighbourhoods by the end of 2024. Overall, this initiative results from an institutionalisation process where RWH has been conceived as a decentralised water management practice by the local authorities.

At the moment of its inception in 2019, *Cosecha de Lluvia* was informed by a study funded by Oxfam and collaboratively prepared by a consortium of academic and non-profit organisations, which estimated the potential of reducing Water Precarity[2] in the city through the implementation of Isla Urbana's systems in single-story houses (Tellman et al., 2019). They found that nearly 40 million litres of water could be captured per year in 21,692 ha of roofs across the city, covering 6%–13% of the city's total water demand (range depending on the storage capacity of a tank from 5,000 to 10,000 L). Other studies (Wunderlich et al., 2021; Concha Larrauri et al., 2019) also

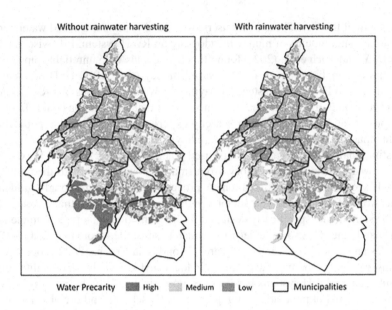

FIGURE 6.1 Estimated change in water precarity after implementing RWH systems, according to Tellman et al. (2019).

show that populations in the southern portion of the city, with higher rates of rain and lower consumption, could cover more than 50% of their water demand for the year. According to Tellman et al. (2019), 415,000 people living in 287 neighbourhoods could significantly reduce their vulnerability if RWH systems are installed in such opportune locations (see Figure 6.1).

So far, *Cosecha de Lluvia* has subsidised more than 60,000 RWH systems in nine boroughs of Mexico City, mainly in the south and east of the city (SEDEMA, 2023).[3] During the period analysed in this review, 2019–2022, the implementation of the programme has evolved as an ongoing learning process. *Cosecha de Lluvia* has been constantly adapted by changing its requirements, target populations, subsidising modes, and operative strategy to benefit more people with the budget available (around 10,000,000 USD per year). For example, the RWH facilities installed were standardised, a gender perspective was incorporated into the objectives and operations of the programme, the follow-up was reinforced to support the adoption of the facilities, and a partial-subsidy modality was incorporated to boost installations in water-stressed but not marginalised neighbourhoods.[4]

Finally, the institutionalisation of RWH in Mexico City has also been supported by legislation. The city's new political constitution, enacted in 2017, mandates the promotion of RWH. Likewise, the Water Law, now called the Law on the Right to Access, Disposal and Sanitation of Water, assigns the city's Environment Secretariat (SEDEMA) the responsibility of installing and operating RWH systems in neighbourhoods that are disconnected from the water network or where water supply is intermittent. Article 16 states that it is the responsibility of Mexico City's water utility (SACMEX) to 'build rainwater collection and storage dams, as well as marginal collectors along ravines and watercourses for water collection', as well as to 'build in ecological reserve zones, green areas, dams, water ponds, infiltration ponds, absorption wells and other works necessary for rainwater collection, in order to increase water levels in the groundwater table, in coordination with the National Water Commission' (Gaceta Oficial del Distrito Federal, 2020). So far, the consolidation of the RWH niche has brought not only changes in law and policy but also joint efforts from a conglomerate of actors involved in the design, commercialisation, and installation of RWH systems, the development and innovation of new devices for RWH facilities, and the study of the technology and its socio-environmental implications.

6.4 SCALING-UP RWH: THE CHALLENGE OF REACHING THE MOST VULNERABLE

The highest levels of water scarcity in Mexico City are expressed in informal settlements (neighbourhoods settled on lands not planned for urban development) along the southern portion of Mexico City. This is a persisting issue since regulations have long prohibited the provision of public services to their residents (Aguilar & López, 2009). Specifically, Article 50 of the 2003 Water Law states that 'water services provided by the authorities may not be provided to those living in irregular (informal) human settlements on conservation land'. This mandate remained in the update, where the law changed its name to 'Law on the Right to Access, Disposal and Sanitation of Water' in 2017 (Hernández Aguilar et al., 2021). Government authorities have blamed informal urban settlements for causing water problems in the city (both flood and water scarcity) by urbanising the conservation land that could be used for aquifer recharge and, as a result, increasing runoff during storms (Eakin et al., 2019; Lerner et al., 2018). Specifically, restrictions on conservation land have been enforced to protect forest areas and guarantee the replenishment of the city's groundwater sources. However, suggesting that informality is the problem, instead of the lack of affordable or public housing options, in part stymies attempts to provide a path to land tenure formalisation and formal urban services such as water supply.

Attempts to 'regularise' informal settlements have been slow, politicised, performative, and rarely occurring on conservation land in Mexico City (Connolly & Wigle, 2017; Wigle, 2020). Government actors, especially in SEDEMA, believe that installing any urban service, such as electricity, piped water, or even RWH systems, would spur continued urban growth over conservation land (Tellman et al., 2021). In the meantime, access to piped water in these areas is minimal, and the government attempts to satisfy needs by sending trucked water. However, the trucked water supply is unreliable, limited, and heavily politicised, often with local governments sending more water to their constituents during elections (de Alba, 2017. Due to the land zoning and conservation regulations prohibiting service provision to informal settlements and the entrenched notion of decision-makers believing that providing water is linked to environmental degradation within SEDEMA, which is the government agency charged with implementing *Cosecha de Lluvia*, it is not surprising that very few RWH systems were installed in the region of the city with the highest water precarity. Note that the term precarity is used instead of scarcity to indicate regions where both the physical conditions of water scarcity and the social conditions of not being able to afford private trucked water storage to solve this problem are present.

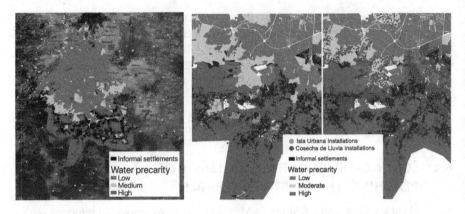

FIGURE 6.2 Left: map of informal settlements as of 2017 in Mexico City (data from SEDEMA, see Tellman et al., 2021 for access) and water precarity. Right: zoom into the southern portion of Mexico City, visualising the locations of RWH systems installed by the city government programme (SEDEMA, from 2019 to 2022) and Isla Urbana installed in 2016–2017 with private funding.

Of the 52,614 systems installed by SEDEMA from 2019 to 2022, only 4,349 systems (about 8%) were installed in informal settlements (analysis by authors, using the SEDEMA's informal settlement data from Tellman et al. (2021) and RWH data from SEDEMA, see Figure 6.2).

Notably, in 2022, the programme rules were adjusted to include 21 informal settlements in the process of regularisation in Tlalpan Borough as beneficiaries. However, excluding informal settlements by large (at least from 2019 to 2022) reflects that the programme has not effectively served the most water-precarious households (see Figure 6.2). Only 3.5% of SEDEMA RWH systems were installed in locations of high water scarcity. 39.8% were installed in medium precarity and 56.8% in low water precarity. By contrast, Isla Urbana, before the onset of government support for RWH systems (using data from 2016 to 2017 installations), installed 7.5% of systems in locations of high water precarity, 50.2% in medium precarity, and 42.3% in locations of low water precarity. The ability to install systems without government regulation regarding informal settlement status enabled Isla Urbana to have a much greater focus on the populations with high and medium levels of water precarity, which arguably has a bigger impact on addressing water scarcity for the most vulnerable in Mexico City. Further, most informal residents are automatically excluded because one of the requirements for getting an RWH system with a full subsidy from *Cosecha de Lluvia* is to present a property tax payment document (*predial* in Spanish). This doubly excluded vulnerable populations that cannot access proper living conditions and basic services for their integral development (Pérez Ortega, 2021).

Despite the restrictions in place, informal residents have found different arrangements to gain access to RWH facilities. Preliminary results from Morelos (2024) found that the dismantling of subsidised RWH systems (when one or more parts of the system are sold, modified, or transformed) is due, in part, to relocating them to informal settlements with less access to water. According to government officials, dismantling remains an isolated problem – about 3% in 2021 (Ibid.). However, many factors influencing the practice and these systems' location and use are not well known yet. Ideally, the 'beneficiaries' should be invited to join participatory processes before, during, and after implementing programmes like *Cosecha de Lluvia*. For instance, Álvarez-Castañón and Tagle-Zamora (2019) showed that in marginalised towns in Guanajuato, Mexico, social programmes are limited to installing eco-technologies, like RWH systems, without effectively involving the people targeted to adopt such facilities. Massively replicating eco-technology projects without understanding socio-natural relationships at local and regional scales could reinforce a technocratic 'fantasy' by focusing only on policies' technical and administrative procedures (Delaney & Fam, 2015). This issue highlights the relevance of building social networks with users (Sofoulis, 2015) and providing them with environmental knowledge of water concerns (Doron et al., 2011) before the system's installations, which is just as, or more, important than the technological implementation itself.

6.5 DISCUSSION: WHAT CAN WE LEARN FROM THIS CASE?

Mexico City has taken significant steps to integrate RWH facilities as a decentralised water management infrastructure. A looming environmental crisis combined with longstanding inequalities in accessing essential services has incentivised the incorporation of RWH technology into the water legal framework and the city's and its boroughs' social policies. Nevertheless, the institutionalisation process documented in this chapter reflects how challenging it is to reach marginalised populations when scaling up the dissemination of technological innovations. RWH has already proven useful in addressing water scarcity, but social and institutional innovations are still required to make it available for the most vulnerable populations. Relegating informal residents, who suffer the most acute water precarity, from accessing RWH facilities results in a double exclusion as they have already been marginalised from proper living conditions and urban services. Ultimately, water scarcity will not be fixed only with technology. It is also necessary to address the underlying conditions explaining why scarcity exists, for whom water is scarce, and why there is unequal access to water across the city in the first place.

Moreover, the increased risks posed by climate-led drought in the Basin of Mexico make RWH a key practice/technology to trigger a sustainable adaptation pathway for the city's water supply. Therefore, RWH systems pose an opportunity to increase water security by enabling the self-sufficiency of households in water access for five to eight months a year and reducing the unpaid time invested in water access. Adopting this technology is particularly beneficial for women in marginalised urban settings, who are usually responsible for obtaining water for their households. As traditional water provisioning systems will not be able to adequately provide for the city's population under current and future climate change conditions, incorporating RWH as an additional water source is vital. While recent droughts have incentivised other governments to adopt this alternative, as in Guadalajara and Monterrey, the second and third most important urban agglomerations in Mexico, new RWH policies should incorporate the learnings of cases like Mexico City, which has been undertaking a process of institutionalisation for more than a decade.

Likewise, enabling proper training and environmental awareness processes are key to ensuring the sustainability of these facilities. As reported by SEDEMA (2020), more than half of the *Cosecha de Lluvia* beneficiaries did not perform the required maintenance in the first year of implementation. While this issue has been addressed over the years, with 85% of the 2022 beneficiaries reporting collecting good-quality rainwater (SEDEMA, 2023), it is important to stress that the social components of RWH policies cannot be overlooked. In practice, social issues like the gender imbalances discussed within the beneficiary households are as important as the technological aspects of RWH.

This chapter does not intend to be an exhaustive review. Many other perspectives are relevant to understanding the cultural, social, political, and environmental shifts associated with incorporating a new, decentralised water management technology, such as RWH, into the current infrastructural regimes in Mexico City. Likewise, many questions remain about the unfolding of this process. Questions like 'How do we achieve the proper adoption of the systems installed in the long term?' or 'How do we maintain the RWH efforts through the coming administrations?' are relevant puzzles to address by furthering the research on this case study. The empirical documentation and theoretical interpretation of current experiences, like those in Mexico City, are essential to building the future of urban RWH in the 21st century.

6.6 CONCLUSION

There is an inherent tension in scaling up RWH from its origins in social movements and private initiatives with government support (like Isla Urbana's efforts before the government programme). Expanding or 'scaling up' innovation using government funds will, by definition, have to obey government regulations and logic. The land use, zoning, environmental laws, and infrastructure regulations in Mexico City are designed to prevent the development of informal settlements. Therefore, it is unsurprising that SEDEMA could not install more RWH systems in informal settlements; it would be against the mandate and mission of their institution to do so. Yet, tackling the water scarcity faced by the most vulnerable populations requires delivering water to informal settlements. While *Cosecha de Lluvia* has installed an important number of RWH systems and is starting to make adjustments to include some informal settlements, the programme's overall impact on the most water-scarce populations is arguably small. To reduce water scarcity for those in the most precarious situation, the city government must either amend its laws to fast-track the regularisation of these settlements or change the programme requirements (as it has already started to do in Tlalpan).

Moreover, it is essential to recognise that technological innovation alone will not fix water scarcity. Legal innovation to change regulations or social innovation by residents or other actors is required. Residents of informal settlements have already started to generate their own strategies to access the programme by having other family members apply. Likewise, continued efforts by Isla Urbana to raise private funds to continue building infrastructure in informal settlements are another alternative. Scaling up RWH for those needing it the most requires more than government mainstreaming. It may involve efforts to change regulations or to make exceptions to environmental laws, as is already starting to happen. According to Fressoli et al. (2013), eco-technology initiatives should be examined in depth so that efforts are not wasted, and the resources and cohesive power of the state are harnessed in comprehensive initiatives. Including the diversity of voices in these kinds of projects implies dealing with resistance to change from different actors and considering and resolving tensions that can lead to conflicts as diverse interests, visions, ideologies, and political cultures of all stakeholders are confronted (Alatorre Frenk et al., 2016).

Addressing the water scarcity problems in Mexico City requires tackling their underlying causes. In other words, addressing why scarcity exists locally, for whom water is scarce, and why there is unequal access to water across the city. Formal property rights are linked to the government's ability to invest in technology to improve water scarcity conditions. Years of effort in urban planning to outline and begin to enact 'regularisation' processes have failed to solve this problem (Connolly & Wigle, 2017; Wigle, 2020). Therefore, addressing why informal urbanisation emerged as a housing solution in the first place, due to the lack of affordable housing for millions of urban residents, could play an important part. Likewise, a stronger involvement of the water utility, SACMEX, could symbolically demonstrate that RWH systems are not just an eco-technology but could play a relevant role as a decentralised water supply infrastructure. These deeper changes must be considered to truly transform the water scarcity panorama and address the inequality in water access that has persisted for decades.

Finally, despite this chapter's focus on *Cosecha de Lluvia* and the importance of addressing the water scarcity of the most vulnerable, it is important to highlight that RWH represents an adaptation measure with a wider potential to contribute to the city's water security. Scaling up the adoption of this technology is also necessary in households that are not currently water-stressed or socially marginalised, as well as in the industry and services sectors. Certainly, addressing the water demand of the populations facing the highest water precarity is a priority. However, climate change is expected to significantly impact critical water sources for the city, such as the Lerma-Cutzamala system, threatening its future viability. Therefore, further policies are required to extend the installation of RWH facilities throughout the city, exploring associations with private entities and other mechanisms rather than government subsidies.

ACKNOWLEDGEMENTS

We thank Professor Omar Masera from the Universidad Nacional Autónoma de México for his views and comments on an earlier version of this chapter.

NOTES

1 Estimated at 0.25MXN/Liter with a range of .05 to .80 depending on the private water provider and borough with large variance (Wunderlich et al., 2021).
2 The study developed a Water Precarity Index, measured by indexing hydrological and social vulnerability variables at a block scale (Tellman et al., 2019).
3 These are Azcapotzalco, Coyoacán, Gustavo A. Madero, Iztapalapa, Magdalena Contreras, Milpa Alta, Tláhuac, Tlalpan, and Xochimilco.
4 Personal communication with Raquel Vargas, technical adviser to the *Cosecha de Lluvia* programme (2024).

REFERENCES

Aguilar, A. G., & López, F. M. 2009. Water Insecurity among the Urban Poor in the Peri-urban Zone of Xochimilco, Mexico City. *Journal of Latin American Geography* 8(2): 97–123. https://www.jstor.org/stable/25765264

Alatorre Frenk, G., Merçon, J., Rosell García, J. A., Bueno García Reyes, I., Ayala-Orozco, B., & Lobato Curiel, V. A. 2016. Para construir lo común entre los diferentes: Guía para la colaboración intersectorial hacia la sustentabilidad. Red de Socioecosistemas y Sustentabilidad y Grupo de Estudios Ambientales, A.C.

Allen, M., Dube, O.P., Solecki, W., Aragón-Durand, F., Cramer, W., Humphreys, S., Kainuma, M., Kala, J., Mahowald, N., Mulugetta, Y. and Perez, R., 2018. Global warming of 1.5 C. An IPCC Special Report on the impacts of global warming of 1.5 C above pre-industrial levels and related global greenhouse gas emission pathways, in the context of strengthening the global response to the threat of climate change, sustainable development, and efforts to eradicate poverty. *Sustainable Development, and Efforts to Eradicate Poverty.*

Álvarez-Castañón, L. del C., & Tagle-Zamora, D. 2019. Transferencia de ecotecnologías y su adopción social en localidades vulnerables: una metodología para valorar su viabilidad. *CienciaUAT* 13(2): 83–99.

Arenas Ortiz. 2021. Sequía 2020–2021: La segunda más severa del registro reciente. Instituto de Ciencias de la Atmósfera y Cambio Climático. https://www.atmosfera.unam.mx/sequia-2020-2021-la-segunda-mas-severa-del-registro-reciente/ (accessed Jan 08, 2024).

Bakir, M., & Xingnan, Z. 2008. GIS and Remote Sensing Applications for Rainwater Harvesting in the Syrian Desert (Al-Badia). *Twelfth International Water Technology Conference IWTC12* 2008: 73–82.

Barkin, D. 2004. Mexico City's Water Crisis. *NACLA Report on the Americas* 38(1): 24–42. https://doi.org/10.1080/10714839.2004.11722401

Bashar, M. Z. I., Karim, Md . R., & Imteaz, M. A. 2018. Reliability and Economic Analysis of Urban Rainwater Harvesting: A Comparative Study within Six Major Cities of Bangladesh. *Resources, Conservation and Recycling* 133: 146–154. https://doi.org/ https://doi.org/10.1016/j.resconrec.2018.01.025

Concha Larrauri, P., Campos Gutierrez, J. P., Lall, U., & Ennenbach, M. 2019. A City Wide Assessment of the Financial Benefits of Rainwater Harvesting in Mexico City. *Journal of the American Water Resources Association* 56(2):1–23. https://doi.org/ 10.1111/1752-1688.12823

Connolly, P., & Wigle, J. 2017. (Re)constructing Informality and "Doing Regularization" in the Conservation Zone of Mexico City. *Planning Theory & Practice* 18(2): 183–201. https://doi.org/10.1080/14649357.2017.1279678

De Alba, F., 2017. Challenging state modernity: Governmental adaptation and informal water politics in Mexico City. *Current Sociology*, *65*(2), pp.182-194.

Delaney, C., & Fam, D. 2015. The 'Meaning' Behind Household Rainwater Use: An Australian Case Study. *Technology in Society* 42: 179–186. https://doi.org/https://doi.org/10.1016/ j.techsoc.2015.05.009

Doron, U., Teh, T.-H., Haklay, M., et al. 2011. Public Engagement with Water Conservation in London. *Water and Environment Journal*, 25(4): 555–562. https://doi.org/https://doi. org/10.1111/j.1747-6593.2011.00256.x

Dumit Gómez, Y., & Teixeira, L. G. 2017. Residential Rainwater Harvesting: Effects of Incentive Policies and Water Consumption over Economic Feasibility. *Resources, Conservation and Recycling* 127: 56–67. https://doi.org/https://doi.org/10.1016/j.resconrec.2017.08.015

Eakin, H., Lerner, A. M., Manuel-Navarrete, D., et al. 2016. Adapting to Risk and Perpetuating Poverty: Household's Strategies for Managing Flood Risk and Water Scarcity in Mexico City. *Environmental Science and Policy* 66: 324–333. https://doi.org/10.1016/ j.envsci.2016.06.006

Eakin, H., Siqueiros-García, J. M., Hernández-Aguilar, B., et al. 2019. Mental Models, Meta-Narratives, and Solution Pathways Associated With Socio-Hydrological Risk and Response in Mexico City. *Frontiers in Sustainable Cities*: 1–13. https://doi.org/10.3389/frsc.2019.00004

Escolero, O., Kralisch, S., Martínez, S. E. et al. 2016. Diagnóstico y análisis de los factores que influyen en la vulnerabilidad del abastecimiento de agua potable a la Ciudad de México, México. *Boletín de La Sociedad Geológica Mexicana* 68(3): 409–427.

Fressoli, M. 2013. Cuando las transferencias tecnológicas fracasan. Aprendizajes y limitaciones en la construcción de Tecnologías para la Inclusión Social. *Universitas Humanística* 76: 73–95.

Gaceta Oficial de la Ciudad de México. 2020. Aviso por el cual se dan a conocer las Reglas de Operación del programa de Sistemas de Captación de Agua de Lluvia en Viviendas de la Ciudad de México. https://www.sedema.cdmx.gob.mx/storage/app/media/DGCPCA/gacetareglas-de-operacion-del-programacosecha-de-lluvia.pdf (accessed Jan 08, 2024).

Gaceta Oficial del Distrito Federal. 2020. Ley del derrecho al acceso, disposición y saneamiento del agua de la Ciudad de México. https://www.sedema.cdmx.gob.mx/storage/app/uploads/public/63b/46c/c97/63b46cc971d5f457815164.pdf (accessed Jan 08, 2024).

García Soler, N., Moss, T., & Papasozomenou, O. 2018. Rain and the City: Pathways to Mainstreaming Rainwater Harvesting in Berlin. *Geoforum* 89: 96–106. https://doi.org/https://doi.org/10.1016/j.geoforum.2018.01.010

Haque, M.M., Rahman, A. and Samali, B., 2016. Evaluation of climate change impacts on rainwater harvesting. *Journal of Cleaner Production*, *137*, pp.60-69.

Hernández Aguilar, B., Lerner, A. M., Manuel-Navarrete, D., et al. 2021. Persisting Narratives Undermine Potential Water Scarcity Solutions for Informal Areas of Mexico City: the Case of Two Settlements in Xochimilco. *Water International* 46(6): 919–937. https://doi.org/10.1080/02508060.2021.1923179

Isla Urbana. 2018 (April 20). *¿Por qué este #DebateChilango fue histórico?* Facebook. https://www.facebook.com/islaurbana/photos/a.121549904550382/1734736789898344/.

Isla Urbana. 2022 (August 11). ¡Alcanzamos los 25 mil sistemas instalados! 💚 👾 #CaptaLaLluvia. Twitter. https://twitter.com/IslaUrbana/status/1557518316532531202.

Karim, Md . R., Bashar, M. Z. I., & Imteaz, M. A. 2015. Reliability and Economic Analysis of Urban Rainwater Harvesting in a Megacity in Bangladesh. *Resources, Conservation and Recycling* 104: 61–67. https://doi.org/10.1016/j.resconrec.2015.09.010

Lee, K. E., Mokhtar, M., Mohd Hanafiah, M. et al. 2016. Rainwater Harvesting as an Alternative Water Resource in Malaysia: Potential, Policies and Development. *Journal of Cleaner Production* 126: 218–222. https://doi.org/10.1016/j.jclepro.2016.03.060

Legorreta, J. 2006. El agua y la Ciudad de México. De Tenochtitlán a la megalópolis del siglo XXI. Universidad Autónoma Metropolitana-Azcapotzalco.

Lerner, A. M., Eakin, H. C., Tellman, E., et al. 2018. Governing the Gaps in Water Governance and Land-Use Planning in a Megacity: The Example of Hydrological Risk in Mexico City. *Cities* 83: 61–70. https://doi.org/10.1016/j.cities.2018.06.009

Li, W., Hai, X., Han, L., et al. 2020. Does Urbanization Intensify Regional Water Scarcity? Evidence and Implications from a Megaregion of China. *Journal of Cleaner Production* 244: 118592. https://doi.org/10.1016/j.jclepro.2019.118592

Martinez, S., Kralisch, S., Escolero, O., et al. 2015. Vulnerability of Mexico City's Water Supply Sources in the Context of Climate Change. *Journal of Water and Climate Change* 6(3): 518–533. https://doi.org/10.2166/wcc.2015.083

McDonald, R. I., Green, P., Balk, D., et al. 2011. Urban Growth, Climate Change, and Freshwater Availability. *Proceedings of the National Academy of Sciences* 108(15): 6312–6317. https://doi.org/10.1073/pnas.1011615108

Mehta, L., 2013. *The limits to scarcity: Contesting the politics of allocation.* Routledge.

Mihelcic, J. R., Zimmerman, J. B., & Ramaswami, A. 2007. Integrating Developed and Developing World Knowledge into Global Discussions and Strategies for Sustainability. *Environmental Science & Technology* 41(10): 3415–3421. https://doi.org/10.1021/es060303e

Morelos, C., M. 2024. Analysis of Social Participation in the Implementation Processes of Public Policy Related to Ecotechnologies: The Case of the Rainwater Harvesting Systems (SCALL) Social Program in Mexico City Homes [Unpublished master's thesis in progress]. Universidad Nacional Autónoma de México.

Mortreux, C., & Barnett, J. 2017. Adaptive Capacity: Exploring the Research Frontier. *WIREs Climate Change* 8(4): e467. https://doi.org/10.1002/wcc.467

Pereira Lindoso, D., Eiró, F., Bursztyn, M., et al. 2018. Harvesting Water for Living with Drought: Insights from the Brazilian Human Coexistence with Semi-Aridity Approach towards Achieving the Sustainable Development Goals. *Sustainability* 10(3): 622. https://doi.org/10.3390/su10030622

Pérez Ortega, V. 2021. *Oportunidades y retos del programa Cosecha de Lluvia de la Ciudad de México, 2020* [Master of Science]. Universidad Nacional Autónoma de México.

Petit-Boix, A., Devkota, J., Phillips, R., et al. 2018. Life Cycle and Hydrologic Modelling of Rainwater Harvesting in Urban Neighborhoods: Implications of Urban Form and Water Demand Patterns in the US and Spain. *Science of the Total Environment* 621: 434–443. https://doi.org/10.1016/j.scitotenv.2017.11.206

Romero-Lankao, P. 2010. Water in Mexico City: What Will Climate Change Bring to its History of Water-Related Hazards and Vulnerabilities? *Environment & Urbanization* 22(1): 157–178. https://www.sedema.cdmx.gob.mx/storage/app/media/CosechaDeLluvia/evaluacion-interna-cdll-2020.pdf

Secretaría del Medio Ambiente SEDEMA. 2020. Evaluación Interna del Programa: Sistemas de Captación de Agua de Lluvia en Viviendas de la Ciudad de México. SEDEMA, Mexico City.

Secretaría del Medio Ambiente SEDEMA. 2021a. *Estrategia Local de Acción Climática 2021–2050/Programa de Acción Climática de La Ciudad de México 2021–2030.* SEDEMA, Mexico City. http://www.data.sedema.cdmx.gob.mx/cambioclimaticocdmx/images/biblioteca_cc/PACCM-y-ELAC_uv.pdf

Secretaría del Medio Ambiente SEDEMA. 2021b. Permite Cosecha de Lluvia que más de 140 mil personas tengan mayor acceso al agua. *Notas.* https://www.sedema.cdmx.gob.mx/comunicacion/nota/permite-cosecha-de-lluvia-que-mas-de-140-mil-personas-tengan-mayor-acceso-al-agua (accessed Jan 08, 2024).

Secretaría del Medio Ambiente SEDEMA. 2023. Cosecha de lluvia. https://www.sedema.cdmx.gob.mx/programas/programa/cosecha-de-lluvia (accessed Aug 25, 2023).

Sivagurunathan, V., Elsawah, S., & Khan, S. J. 2022. Scenarios for Urban Water Management Futures: A Systematic Review. *Water Research*, 211: 118079. https://doi.org/10.1016/j.watres.2022.118079

Smith, A. 2005. The Alternative Technology Movement: An Analysis of its Framing and Negotiation of Technology Development. *Human Ecology Review* 12(2): 106–119.

Sofoulis, Z. 2015. The Trouble with Tanks: Unsettling Dominant Australian Urban Water Management Paradigms. *Local Environment* 20(5): 529–547. https://doi.org/10.1080/13549839.2014.903912

Tellman, B., de Alba, F., Serrano-Candela, F., et al. 2019. Captación de lluvia en la CDMX : Un análisis de las desigualdades espaciales. https://www.oxfammexico.org/sites/default/files/Captacionde agua en la CDMX.pdf (accessed Jan 08, 2024).

Tellman, B., Eakin, H., Janssen, M. A., et al. 2021. The Role of Institutional Entrepreneurs and Informal Land Transactions in Mexico City's Urban Expansion. *World Development* 140: 105374. https://doi.org/10.1016/j.worlddev.2020.105374

Torres Bernardino, L. 2017. La gestión del agua potable en la Ciudad de México: los retos hídricos de la CDMX: gobernanza y sustentabilidad. Instituto Nacional de Administración Pública, A.C.

United Nations (UN). 2022. The United Nations World Water Development Report 2022: Groundwater: Making the Invisible Visible. UNESCO, Paris.

Wigle, J. 2020. Fast-Track Redevelopment and Slow-Track Regularization: The Uneven Geographies of Spatial Regulation in Mexico City. *Latin American Perspectives* 47(6): 56–76. https://doi.org/10.1177/0094582X19898199

WHO/ UNICEF. 2022. Joint Monitoring Programme (JMP) for Water Supply, Sanitation and Hygiene https://washdata.org/data/household#!/table?geo0=country&geo1=MEX (accessed Jan 08, 2024).

Wunderlich, S., George Freeman, S., Galindo, L., et al. 2021. Optimizing Household Water Decisions for Managing Intermittent Water Supply in Mexico City. *Environmental Science & Technology* 55(12): 8371–8381. https://doi.org/10.1021/acs.est.0c08390

Zheng, C., Liu, J., Cao, G., et al. 2010. Can China Cope with Its Water Crisis? Perspectives from the North China Plain. *Groundwater* 48(3): 350–354. https://doi.org/10.1111/j.1745-6584.2010.00695_3.x

7 The Role of Boardwalk Cistern in Supplementing Water-deficient Irrigation in a Semiarid Region

Juliana Farias Araujo,
Anderson de Souza Matos Gadéa,
Ana Caroline Bastos Lima de Souza,
Maria Auxiliadora Freitas dos Santos,
and Eduardo Borges Cohim

7.1 INTRODUCTION

Semiarid regions can be characterized by an aridity index (AI) between 0.20 and 0.50 and precipitation within the range of 200 to 700 mm a year (Lal 2004; Gallart et al. 2002). They also usually have an evaporation/precipitation ratio greater than 1.0, which contributes to the water deficit in the region (Lu et al. 2018).

In Brazil, semiarid regions are classified in a broader way, presenting at least one of the following characteristics: average precipitation equal to or less than 800 mm per year; AI equals to or less than 0.50; or a daily water deficit equal to or higher than 60%, considering all days of the year (Brasil 2021). This restriction has a negative impact on regional agricultural production, mainly done by smallholder and family farmers, who are dependent on climate conditions and manual labour since they do not have easy access to irrigation infrastructure and technical support (Sampaio et al. 2017; Singhal et al. 2020).

As a result, the majority of the 1.5 million families living in the semiarid region are among the poorest in Brazil, with an average Human Development Index (HDI) rarely reaching 0.5 (Melo 2017). Considering the impossibility of altering these regional climate characteristics, policies of "Coexistence with the Semiarid Region" appear as an alternative to make sustainable family farming viable. This paradigm seeks to promote the monitoring, adaptation, and mitigation of the conditions imposed by the climate while valuing and conserving the potential of local natural resources (Brito et al. 2017).

In Brazil, the "One land and two waters" programme (P1 + 2) aligns with this paradigm by encouraging the construction of rainwater harvesting systems (RWHS) – in the broad sense – such as underground dams, microdams, stone tanks and boardwalk cisterns. The latter consists of a cistern with a standard volume of $52\,m^3$ that stores

DOI: 10.1201/9781032638102-11

rainwater collected on a paved boardwalk catchment surface of $200\,m^2$ in area. The water collected in this system has good quality and can be used during the dry season to meet the water demands of family farms for crop and livestock production (Sánchez et al. 2015).

Related literature has shown that such a system is well accepted by users, mainly due to the increase in farm production, and even better results have been reported when in association with irrigation systems (Cavalacante et al. 2021; Silva et al. 2021). In this way, encouraging irrigation techniques is important, such as supplementing water-deficient irrigation by employing "water depths" that maximise water productivity instead of crop productivity.

It is noteworthy that research on the potential of RWHS as a source for irrigation usually considers meeting crop total demands (Gris et al. 2017; Jing et al. 2017; Minatto 2013) instead of optimizing the use of water. The latter maximise crop yields per volume of water applied, as is necessary for semiarid regions. In this context, model simulation can be used to plan and supplement water-deficient irrigation by comparing scenarios, optimizing resource management for agricultural production, considering crop specificities and hydrological characteristics, and varying water input (Wang et al. 2022).

This work aimed to propose a framework to assess the service level of rainwater harvesting using boardwalk cisterns to meet irrigation demands in the semiarid region, with a focus on water productivity. The framework was then evaluated in a study case in Feira de Santana, Brazil.

7.2 METHODS

The service level of an RWHS was verified based on its service efficiency (SE), which indicates the portion of the demand from irrigation that is supplied daily by the stored rainwater. In this study, the proposed framework to calculate SE followed the following steps (Figure 7.1): (i) collect meteorological data for the region of study (precipitation, temperature, evapotranspiration); (ii) define the crop and model its root depth and growth; (iii) based on meteorological data and crop growth models, calculate total irrigation depths that maximise crop yield, generating a time series of daily demand for each irrigation depth; (iv) define a time series that maximises water productivity; and (v) calculate the SE using the water balance model proposed by Fewkes (2000). This method can be applied to any location, for different crops, and to all combinations of catchment area and cistern volume.

7.2.1 WATER DEMAND FOR IRRIGATION

The daily water demand for irrigation was estimated by using a tool called Global Evolutionary Technique for OPTimal Irrigation Scheduling (GET-OPTIS) (Schütze and Mialyk 2019), included in the Deficit Irrigation Toolbox (DIT) software. This tool applies evolutionary optimization algorithms to search for optimal schedules and irrigation depths, generating a series of daily irrigation demands that maximise production (Schütze et al. 2011).

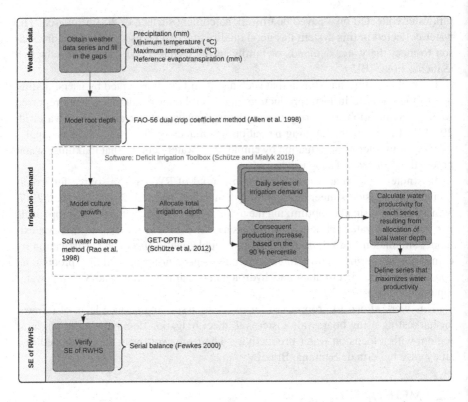

FIGURE 7.1   Framework to calculate the service efficiency of a rainwater harvesting system for irrigation.

Input data for the GET-OPTIS includes the crop root depth model, which is used to calculate the soil water balance (Rao et al. 1998). In this study, crop root depth was modelled using the FAO-56 dual crop coefficient method (Allen et al. 1998).

Crop productivity was calculated for several irrigation depths, varying from 0 mm to the depth that corresponds to full supplemental irrigation for that crop, generating a series of daily demands for each total depth. The irrigation depths used for the service efficiency assessment are chosen based on the maximum water productivity, calculated as in Eq. (7.1).

$$WP = \frac{CP}{P + Iv} \tag{7.1}$$

where WP is the water productivity (kg·m³), CP is the crop production (kg), P is the total precipitation (m³) and Iv is the irrigation volume (m³).

7.2.2   Rainwater Harvesting System Service Efficiency

The water use for the RWHS through time was calculated as in Eqs. (7.2, 7.3, and 7.4) (Fewkes 2000).

$$Q_{(t)} = P_{(t)} \cdot A \cdot C \qquad (7.2)$$

$$Y_{(t)} = \min \begin{cases} D_{(t)} \\ V_{(t-1)} + \theta Q_{(t)} \end{cases} \qquad (7.3)$$

$$Y_{(t)} = \min \begin{cases} V_{(t-1)} + Q_{(t)} - Y_{(t)} \\ R - (1-\theta)Y_{(t)} \end{cases} \qquad (7.4)$$

where $Q_{(t)}$ is the daily waterflow to the cistern (L), $P_{(t)}$ is the daily precipitation (mm), A is the catchment area (m²), C is the runoff coefficient (assumed as 0.8), $Y_{(t)}$ is the water used to supply the daily demand, θ is the dimensionless coefficient that varies from 0 to 1 (assumed as 0.5 for an average situation), $D_{(t)}$ is the daily demand, $V_{(t)}$ is the volume of water stored (L) and R is the cistern capacity (L).

The daily demand for irrigation was calculated as in Eq. (7.5).

$$D_{(t)} = \frac{0.4 \cdot Iv_{(t)}}{Efi} \qquad (7.5)$$

where Efi is the irrigation efficiency, adopted as 0.9 for dripping irrigation (Ferreira 2011), and the constant 0.4 represents the minimum fraction of wet area recommended for semiarid conditions (EMBRAPA 2009).

The SE (%) of RWHS can be calculated as in Eq. (7.6):

$$SE = \frac{\sum Y_{(t)}}{\sum D_{(t)}} 100 \qquad (7.6)$$

7.2.3 CASE STUDY: TOMATO CROP IN FEIRA DE SANTANA

The proposed framework was applied to a tomato production system with a planting area of 2,500 m² in the rural area of Feira de Santana, State of Bahia, Brazil, located in a semiarid region. Planting periods for the main crops in the region usually begin between May and July, which are the months with greater precipitation rates and lower temperatures (Santos et al. 2018). Therefore, the planting date was simulated to begin on the first day of May.

7.2.3.1 Data Sources

The data series for daily temperature (minimum and maximum), precipitation, and reference evapotranspiration (ET0) were obtained from the National Institute for Meteorology (INMET 2018) for a period of 20 years, from 1998 to 2017. Gaps in the ET0 data set were filled using AquaCrop (FAO 2021), and gaps in temperature and precipitation datasets were filled using data from Xavier, King and Scalon (2016). The complete dataset presents average values of 705 mm for annual precipitation, 1,159 mm for annual potential evapotranspiration, as well as 20°C and 30°C for minimal and maximum annual temperatures.

7.2.3.2   Crop Growth and Production Modelling

Literature data were used to estimate the parameters of the root depth model and calibrate the DIT software (Table 7.1). A planting area of $2,500\,m^2$ was adopted, and eight total irrigation depths were used (20, 40, 60, 80, 100, 150, 200 and 250) mm per m^2, considering a scenario of supplementing water-deficient irrigation. Two other scenarios were simulated: no irrigation, i.e. rainfed production – precipitation as the only source of water for crops – and full supplemental irrigation – assuming unlimited water availability for irrigation and application of water whenever soil moisture is lower than 90% of field capacity.

7.2.3.3   Assessing Rainwater Harvesting Efficiency

The RWHS adopted for the case study was a boardwalk cistern with a storage capacity of $52\,m^3$ and a catchment area of $200\,m^2$ (Figure 7.2). These dimensions are commonly used within the P1 + 2 programme, which aims to promote actions towards the coexistence with semiarid climate conditions, mainly by building RWHS in rural households and farms to meet human, crop, and livestock water demands.

Considering that the SE reflects the global performance of the RWHS during the simulated period, the performance of the system for each year of the series was also analysed, taking into account the consecutive dry days (CDD) throughout the year, the number of rainfall days (NRD), and the coefficient of variation of rainfall (CV). Days with any rainfall were considered to be those in which precipitation was equal to or greater than 1 mm. The rainfall anomaly index (RAI) (Van-Rooy 1965) was calculated as in Eq. (7.7) (positive anomalies) and Eq. (7.8) (negative anomalies), and the classification proposed by Araújo, Moraes Neto and Sousa (2009) was used to characterise the severity of the dry and rainy seasons based on the RAI values.

$$RAI^{(+)} = 3 \left(\frac{N - \bar{N}}{M - \bar{N}} \right) \tag{7.7}$$

$$RAI^{(-)} = -3 \left(\frac{N - \bar{N}}{\bar{X} - \bar{N}} \right) \tag{7.8}$$

TABLE 7.1

Data Used to Model Root Depth and Calibrate the DIT Software

Data	Value	Unit	Reference
Maximum plant height	0.6	m	Allen et al. (1998)
Field capacity and permanent wilting point (sandy-loam texture)	22 %; 10 %	-	Raes (2017)
Initial and mean basal coefficient	0.15; 1.1	-	Allen et al. (1998)
Crop coefficients	0.40; 0.75; 1.11; 0.93	-	Freitas (2018)
Phenological phases durations	6; 26; 33; 24	day	Freitas (2018)
Maximum productivity	55	t·ha[1]	Doorenbos and Kassam (1979)
Seeding depth	4	cm	Freitas (2018)
Maximum effective depth	1.1	m	Allen et al. (1998)

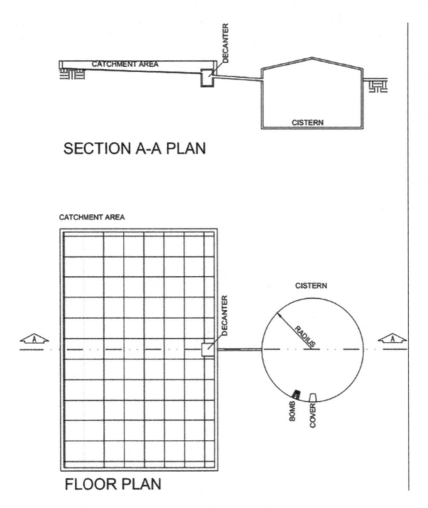

FIGURE 7.2   Graphical representation of a boardwalk cistern. Source: Giffoni et al. (2019).

where N is the annual precipitation (mm), $\overline{N}$ is the average annual precipitation of the entire historical series (mm), $\overline{X}$ is the average annual precipitation of the ten years with the smallest precipitation (mm), and M is the average annual precipitation of the ten years with the largest precipitation (mm).

7.3   RESULTS AND DISCUSSION

Simulations showed a maximum production of 6 tonnes of tomatoes considering a 2,500 m² planted area with no irrigation (water from precipitation only). Figure 7.3 shows the results for increments in tomato production varying the series of daily irrigation demands resulting from the temporal and quantitative allocation of the total irrigation depths compared with the rainfed scenario.

The incremental yields tend to decrease as irrigation depths increase. The same pattern was seen by Orduña-Alegria et al. (2019) when analysing corn crop irrigation

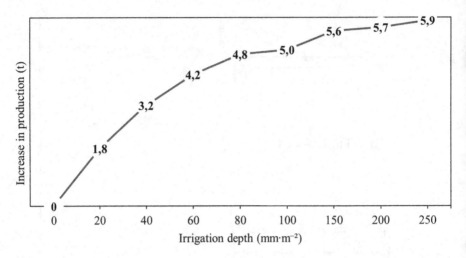

FIGURE 7.3   Increase in tomato crop yield by the raise in the supplement to water-deficient irrigation compared with a non-irrigated system.

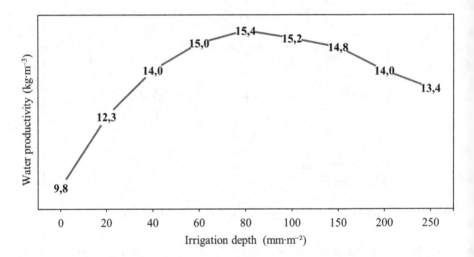

FIGURE 7.4   Water productivity by water-deficient irrigation supplementation.

in the United States. Therefore, water productivity (Figure 7.4) was used as the choice criteria for the irrigation rate employed in the assessment of the SE of the boardwalk cistern, seeking always to maximise water instead of crop productivity.

The simulated crop production from the series of daily irrigation demand from the allocation of a depth of 80 mm/m² was the one that maximised water productivity (15.4 kg/m³), with an increment in the production of 4.8 tonnes (56%) when compared with the system with no irrigation.

The scenario with unlimited access to water for irrigation presented the maximum production when depth was 221 mm/m², resulting in an additional 7.3 tonnes of yield compared with the non-irrigated system. However, this would represent $157\,m^3$ more water demand than the water-deficient irrigation scenario, which is critical for the agriculture sector in the semiarid region.

The water saved in the water-deficient irrigation scheme ($157\,m^3/m^2$) could be used to vary crop production or to meet the annual water demand of 172 goats, considering a consumption of 5 L·head^{-1}·day^{-1} and a six-month growth period (IABS 2011). Water savings could be even greater, as in the experimental study in China by Sun et al. (2022), whose findings showed that increasing apple production by 550 tonnes per·year in an area of $1.22 \times 10^7\,m^2$ can save around $15,000\,m^3$ of water and increase water productivity by 33.4%.

Although the supplementation of water-deficient irrigation enables reduced water use, it does not guarantee to meet water demand for the entire year due to the limited availability of water during the dry season, the concentration of rainfall in 1/3 of the year and the low water soil storage capacity (Andrade et al. 2017).

In this context, the boardwalk cistern stands as a major solution to store rainwater during the dry season, minimizing transpiration loss due to the cover sealing. In addition, its combined use with irrigation systems has the potential to minimise soil salinization by storing low saltwater, contrary to open tank storage.

The boardwalk cistern from the P1 + 2 program assessed herein showed a SE of 92% over the simulated 20-year period for the series of daily irrigation demand from a total depth of $80\,mm/m^2$. High SE values were also observed in the annual analysis (Table 7.2), remaining above 70% even for years considered extremely dry (e.g. 2017). For instance, this is in accordance with the Chinese Ministry of Agriculture recommendations, which state a minimum guarantee within the range of 50% to 75% (Zhu et al. 2015).

These findings corroborate the RWHS's potential to significantly improve the living conditions of family farmers in semiarid climates. For instance, for an increase in crop yield of 4.8 tonnes and a market price of R$ 4.30 per kilogram of tomato, there would be a gross revenue increase of R$ 21,000 (about USD 4,350, using the May 2022 exchange rate) compared with the non-irrigated system. This increase is, however, dependent on the type of crop. Planting and irrigation planning must be conducted to maximise the benefits obtained by the RWHS, as well as the use of organic fertilizers and natural pesticides.

Identifying the "best use of stored water" could further enhance these benefits. A surplus of water during the planting period (March–June) was verified (Figure 7.5). which could be used for livestock production, especially small animals such as goats and chickens, or for growing vegetables.

Another point to note is that, for the case study herein, it would be possible to reach a SE of 90% even with a cistern storage capacity of $11.1\,m^3$, which is substantially lower than the $52\,m^3$ adopted by the P1 + 2 program. Therefore, despite the need for standardization in public policies to some degree, designing RWHS for regional conditions and local needs would result in lower costs and better use of materials.

Employing such systems to address water scarcity in semiarid climates may also reduce the demand for centralised water supply infrastructure and its associated environmental impacts and minimize groundwater and surface water exploitation. This becomes critical in the context of a decrease in the flooded surface area of the Caatinga (predominant biome) by 8.3%, i.e. around $800\,km^2$. This scenario tends to

TABLE 7.2
Precipitation Parameters and Service Efficiency of the Boardwalk Cisterns in the Feira de Santana Region

Year	P (mm)	RAI	RAI classification	NRD	CV	CDD	SE (%)
1999	882	5.15	Extremely rainy	107	2.8	28	96.7
2000	874	4.91	Extremely rainy	110	2.9	28	100
2010	851	4.23	Extremely rainy	106	2.8	27	86.9
2003	824	3.45	Very rainy	122	2.4	21	100
2008	792	2.52	Very rainy	105	3.1	30	89.2
2011	788	2.40	Very rainy	87	3.7	26	80.1
2005	784	2.29	Very rainy	107	3.4	30	100
2004	774	2.01	Very rainy	71	3.6	40	90.2
2007	772	1.94	Rainy	103	3.6	21	100
2006	743	1.09	Rainy	95	2.8	48	97.1
2013	732	0.78	Rainy	104	3.0	38	91.8
1998	711	0.19	Rainy	109	2.7	18	80.0
2014	670	−1.00	Dry	112	2.7	18	96.0
2015	649	−1.64	Dry	99	3.1	43	100
2002	645	−1.75	Dry	87	3.1	32	97.3
2001	619	−2.49	Very dry	88	2.6	35	87.7
2016	608	−2.81	Very dry	83	3.4	26	86.3
2009	580	−3.64	Very dry	73	3.2	29	100
2017	429	−8.01	Extremely dry	92	2.5	58	70.5
2012	373	−9.64	Extremely dry	82	3.5	25	84.4

P – Mean annual precipitation, RAI – rainfall anomaly index, NRD – number of rain days, CV – coefficient of variation, CDD – consecutive dry days.

worsen as this semiarid region has been gaining prominence for food production, with projected growth in irrigated areas of more than 76%, i.e. about 42,000 km^2 (ANA 2021; MapBiomas 2021).

Providing RHWS for all the 1.5 million families residing in the semiarid region would reduce the water footprint by 64% and increase the production area by 360,000 hectares, which corresponds to 3.5 times the public irrigated area and is around twice as large as the projected expansion in irrigated areas of nearly 2,000 km^2 (ANA 2021). This could further encourage sustainable agricultural production. Thus, shifting away from unsustainable activities such as firewood and charcoal production, which is the main cause of deforestation of the Caatinga native vegetation that consumed around 100,000 km^2 from 1985 to 2020 (MapBiomas 2021).

Furthermore, food production with such systems would not require any additional surface water use (blue water), collaborating with the maintenance of ecological flows and preserving water and ecosystem services. Also, energy demand would be minimised by decentralised production, a lower distance for water transport, and a virtual absence of barriers.

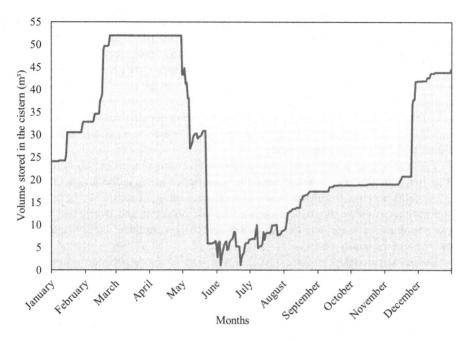

FIGURE 7.5   Simulation of the volume of water available in the cistern considering irrigation demands.

From a social point of view, the increase in the productive area would directly benefit the food and water security of family farmers and allow for higher diversity in production by enabling them to plant long-cycle crops without being limited by rainy seasons. It would also reduce issues associated with the rural exodus, as the additional access to water would lessen the need to buy food and might as well increase their income due to the sale of surplus production in nearby markets. In this sense, it would provide the achievement of the Sustainable Development Goals (SDG) proposed by the United Nations (UN), especially with regard to food sovereignty (SDG 2) and increasing the income of these farmers (SDG 10) (UN 2019).

The findings from Fagundes et al. (2020) support these potential benefits, as 97% of family farmers benefiting from the P1 + 2 programme reported improvements in the quality and diversity of food produced. In addition, 72.4% confirmed the increase in income, which shows how the sidewalk cisterns allowed the promotion of food sovereignty, preserving traditional production and food practices.

7.4   CONCLUSION

This study proposed a methodology to assess the role of boardwalk cisterns in meeting the supplementary demand for water-deficient irrigation. The quantitative and temporal allocation of the total irrigation depth and its productivity were evaluated to identify the best use of water at each stage.

The case study with tomato crops in Feira de Santana, located in the Brazilian semiarid region, showed that the daily series application of irrigation demand from the allocation of an 80 mm total depth resulted in the maximum productivity of water (15.4 kg/m^3) and an increase in production of 4.8 tonnes (80%) for a 2,500 m^2 area when compared to the rainfed condition. These 4.8 tonnes are 32% lower than the production that would be obtained with the total supplementary irrigation, which would require 177% more water. The deficit irrigation demand could be met – with a global service efficiency of 92% – over the next 20 years by a system with a 52 m^3 cistern and 200 m^2 catchment area.

The widespread adoption of such a system on small properties throughout the Brazilian semiarid region would have the potential to increase the production area by 360,000 hectares, resulting in social and water usage benefits. Further positive implications would be the reduced demand for infrastructure and energy to transport a large volume of water over long distances and the generation of more occupations and income for family farmers.

The growth model for soil-water balance was used herein. As accurate as this model can get, it does not consider the occurrence of pests, which could influence production. Therefore, it is recommended for future work to conduct experiments to adjust the parameters as well as to prove the simulated results. In addition, it is also recommended to verify the possibility of minimizing construction costs and enhancing the benefits arising from cisterns that demand more water in the RWHS design phase.

ACKNOWLEDGMENTS

The authors thank the financial support of the Bahia State Research Support Foundation (FAPESB) for J. F. Araujo (No. 1204/2020) and the National Council for Scientific and Technological Development (CNPq) for C. B. L. Sousa (No. 159878/2020–9). The authors thank Samuel Alex Sipert and Adriano Souza Leão for their contributions to the revision of the English language and editing.

REFERENCES

Agência Nacional de Águas e Saneamento Básico - ANA. 2021. *Atlas Irrigação 2021: uso da água na agricultura irrigada.* Brasília: ANA.
Allen, R. G., Pereira, L. S., Raes, D., & Smith, M. 1998. *Crop Evapotranpiration: Guidelines for computing crop water requirements.* Rome: FAO.
Andrade, E. M. de, Aquino, D. do N., Chaves, L. C. G., & Lopes, F. B. 2017. Water as capital and its uses in the Caatinga. In: *Caatinga: the largest tropical dry forest region in South America*, ed. J. M. C. da Silva, I. R. Leal, and M. Tabarelli, 281–302. [S.L.]: Spring.
Araújo, L. E., Moraes Neto, J. M., & Sousa, F. A. S. 2009. Classification of annual rainfall and the rainy quarter of the year in the Paraíba river basin using rain anomaly index (RAI). *Revista Ambiente e Água*, no. 4: 93–110. https://doi.org/10.4136/ambi-agua.105
Brasil. Conselho Deliberativo da Superintendência do Desenvolvimento do Nordeste. 2021. Resolução nº 150, de 13 de dezembro de 2021. Aprova a Proposição n. 151/2021, que trata do Relatório Técnico que apresenta os resultados da revisão da delimitação do Semiárido 2021, inclusive os critérios técnicos e científicos, a relação de municípios habilitados, e da regra de transição para municípios excluídos. Diário Oficial [da] República Federativa do Brasil, Brasília.

Brito, F. C. da S., Lima, D. C, & Sousa, J. D. 2017. Uma abordagem histórica e teórica das políticas públicas de combate à seca e convivência com o semiárido. *Revista Brasileira de Gestão Ambiental*, no. 11, 57–65.

Cavalacante, L., Mesquita, P. S., & Rodrigues-Filho, S. 2021. Cisternas de 2a Água: tecnologias sociais promovendo capacitação adaptativa às famílias de agricultores brasileiros. In: *A ação pública de adaptação da agricultura à mudança climática no Nordeste semiárido brasileiro*, ed. E. Sabourin, L. M. R. O. F. Goulet, and E. S. Martins, 123–142. Rio de Janeiro: E-Papers Serviços Editoriais Ltda.

Doorenbos, J., & Kassam, A. H. 1979. Yield response to water. Rome: FAO.

Empresa Brasileira De Pesquisa Agropecuária - EMBRAPA. 2009. Sistema de Produção de Melancia. https://sistemasdeproducao.cnptia.embrapa.br/FontesHTML/Melancia/SistemaProducaoMelancia/irrigacao.htm

Fagundes, A. A., Silva, T. C., Voci, S. M., dos Santos, F., Barbosa, K. B. F., & Corrêa, A. M. S. 2020. Food and nutritional security of semi-arid farm families benefiting from rainwater collection equipment in Brazil. *PLoS One*, no. 15, 1–14. https://dx.doi.org/10.1371/journal.pone.0234974

Ferreira, V. M. 2011. Irrigação e Drenagem. Floriano: EDUFPI.

Fewkes, A. 2000. Modelling the performance of rainwater collection systems: towards a generalised approach. *Urban Water*, no. 1, 323–333. https://doi.org/10.1016/S1462-0758(00)00026-1

Freitas, J. C. de. 2018. Calibração do modelo Aquacrop e necessidades hídricas da cultura do tomateiro cultivada em condições tropicais (Doctoral thesis). Center for Technology and Natural Resources, Graduate Program in Meteorology, Federal University of Campina Grande, Campina Grande.

Gallart, F.; Solé, A.; Puigdefàbregas, J., & Lázaro, R. 2002. Badland systems in the Mediterranean. In *Dryland rivers: hydrology and geomorphology of semi-arid channels*, ed. Bull, L. J., and Kirkby, M. J., 299–326. Chichester: Wiley.

Giffoni, V. V., Gadéa, A. S. M., Cohim E., Freitas, J. J., & Araujo, J. F. 2019. Sizing of rainwater haversting systems for animal watering in semiarid region. *Water Practice and Technology*, no. 14, 971–980. https://doi.org/10.2166/wpt.2019.080

Gris, V. G. C., Bertolini, G. R. F., & Johann, J. A. 2017. Rural tanks: economic viability and perception of farmers in Palotina-PR. *Revista Nera*, no. 37, 169–194. https://doi.org/10.47946/rnera.v0i37.4755

Instituto Ambiental Brasil Sustentável - IABS. 2011. Manual de criação de caprinos e ovinos. Brasília: CODEVASF.

Jing, X, Zhang, S., Zhang, J, Wang, Y., & Wang, Y. 2017. Assessing efficiency and economic viability of rainwater harvesting systems for meeting non-potable water demands in four climatic zones of China. *Resources, Conservation and Recycling*, no. 126, 74–85. https://doi.org/10.1016/j.resconrec.2017.07.027

Lal, R. 2004. Soil carbon sequestration impacts on global climate change and food. *Security Science*, no. 304, 1623–1627. https://dx.doi.org/10.1126/science.1097396

Lu, C., Zhao, T., Shi, X., & Cao, S. 2018. Ecological restoration by afforestation may increase groundwater depth and create potentially large ecological and water opportunity costs in arid and semiarid China. *Journal of Cleaner Production*, no. 176, 1213–1222. https://dx.doi.org/10.1016/j.jclepro.2016.03.046.

Melo, F. P. L. 2017. The Socio-Ecology of the Caatinga: Understanding How Natural Resource Use Shapes an Ecosystem. In *Caatinga: the largest tropical dry forest region in South America*, ed. J. M. C. da Silva, I. R. Leal, and M. Tabarelli, 281–302. [S.L.]: Spring.

Minatto, M. M. 2013. Água de chuva: uso para irrigação em agricultura familiar. (Monography). Faculty of Civil Engineering, Federal University of Rio Grande do Sul, Porto Alegre.

National Institute for Meteorology - INMET. 2018. Weather Data for Feira de Santana. https://portal.inmet.gov.br/

Organização para a Alimentação e Agricultura - FAO. 2021. AquaCrop. https://www.fao.org/
    land-water/databases-and-software/aquacrop/en/
Orduña-Alegria, M. E., Schütze, N., & Niyogi, D. 2019. Evaluation of hydroclimatic variabil-
    ity and prospective irrigation strategies in the U. S. Corn Belt. *Water*, no. 11. https://doi.
    org/10.3390/w11122447
Projeto de Mapeamento Anual do Uso e Cobertura da Terra no Brasil - MapBiomas. 2021.
    Mapeamento anual da cobertura e uso da terra no Brasil (1985-2020): destaques caat-
    inga. [S. L]: Mapbiomas.
Raes, D. 2017. AquaCrop training handbooks I: understanding Aquacrop. Roma: FAO.
Rao, N. H., Sarma, P. B. S., & Chander, S. 1998. A simple dated water-production function for
    use in irrigated agriculture. *Agricultural Water Management*, no. 13, 25–32. https://doi.
    org/10.1016/0378-3774(88)90130-8
Sampaio, E. V. de S. B., Menezes, R. S. C., Sampaio, Y. de S. B., & de Freitas, A. D. S. 2017.
    Sustainable Agricultural Uses in the Caatinga. In *Caatinga: the largest tropical dry for-
    est region in South America*, ed. J. M. C. da Silva, I. R. Leal, and M. Tabarelli, 281–302.
    [S.L.]: Spring.
Sánchez, A. S., Cohim, E., & Kalid, R. A. A. 2015. Review on physicochemical and microbio-
    logical contamination of roof-harvested rainwater in urban areas. *Sustainability of Water
    Quality and Ecology*, no. 6, 119–137. https://doi.org/10.1016/j.swaqe.2015.04.002
Santos, R. A. dos, Martins, D. L., & Santos, R. L. 2018. Water balance and Köppen climate
    classification and thornthwaite in the municipality of Feira de Santana (BA). *Geo Uerj*,
    no. 33, 1–17. https://doi.org/10.12957/geouerj.2018.34159
Schütze, N., & Mialyk, O. 2019. *Deficit irrigation toolbox (DIT): user guide*. Dresden,
    Germany .
Schü tze, N., de Paly, M., & Shamir, U. 2011. Novel simulation-based algorithms for opti-
    mal open-loop and closed-loop scheduling of deficit irrigation systems. *Journal of
    Hydroinformatics*, no. 14, 136–151. https://doi.org/10.2166/hydro.2011.073
Silva, T. A., Ferreira, J., Calijuri, M. L., dos Santos, V. J., do Alves, S. C., & de Castro, J.
    S. 2021. Efficiency of technologies to live with drought in agricultural development
    in Brazil's semi-arid regions. *Journal of Arid Environments*, no. 192, 1–13. https://doi.
    org/10.1016/j.jaridenv.2021.104538
Singhal, A., Gupta, R., Singh, A. N., & Shrinivas, A. 2020. Assessment and monitoring of
    groundwater quality in semi -arid region. *Groundwater for Sustainable Development*,
    no. 11. https://doi.org/10.1016/j.gsd.2020.100381
Sun, M., Gao, X., Zhang, Y., Song, X., & Zhao, X. 2022. A new solution of high-efficiency
    rainwater irrigation mode for water management in apple plantation: design and
    application. *Agricultural Water Management*, no. 259. https://dx.doi.org/10.1016/j.
    agwat.2021.107243
United Nations Organization - UN. 2019. Sustainable development goals. https://www.un.org/
    sustainabledevelopment/wp-content/uploads/2019/01/SDG_Guidelines_AUG_2019_
    Final.pdf
Van-Rooy, M. P. 1965. A rainfall anomaly index (RAI) independent of time and space. *Notos*,
    no.14, 43–48.
Wang, Y., Guo, S. S., & Guo, P. 2022. Crop-growth-based spatially-distributed optimiza-
    tion model for irrigation water resource management under uncertainties and future
    climate change. *Journal of Cleaner Production*, no. 345. https://doi.org/10.1016/j.
    jclepro.2022.131182
Xavier, A. C., King, C. W., & Scanlon, B. R. 2016. Daily gridded meteorological variables in
    Brazil (1980–2013). *International Journal of Climatology*, no. 36, 2644–2659. https://
    doi.org/10.1002/joc.4518
Zhu, Q., Gould, J., Li, Y., & Ma, C. 2015. *Rainwater harvesting for agriculture and water
    supply*. Londres: Spring.

8 Designing School Rainwater Harvesting Systems in Water-Scarce Developing Countries

Jeremy Gibberd

8.1 INTRODUCTION

Hotter temperatures, longer dry spells and an increasing number of droughts have led to water scarcity in many areas of the world (Diedhiou et al. 2018; Makki et al. 2015; IPCC 2022). This situation is getting worse, and UNESCO World Water Assessment Programme (2019) projects that two-thirds of the world's population will experience water scarcity for at least one month of the year, and there will be a 40% gap between water demand and supply by 2030. In developing countries, this situation is aggravated by rapid urbanisation and a lack of capacity and resources to scale up existing water supply systems (UN-Habitat and IHS-Erasmus University Rotterdam 2018). Existing infrastructure that is already struggling to meet demand may also not be adequately maintained, resulting in high levels of leakage, breakdowns, and an increasingly unreliable water supply (Wensley and Mackintosh 2015).

Eighty-eight percent of deaths caused by diarrheal disease are attributed to a lack of water and inadequate sanitation (World Health Organization 2009). Nearly all these deaths are in developing countries, and 84% are children. Inadequate water supplies and poor sanitation are the main causes of child mortality, and it is estimated that 443 million school days are lost per year from water-related illnesses (Watkins, 2006). Schools need water for drinking, cleaning, and flushing toilets (if they have water-borne sanitation) (Jasper et al. 2012). A lack of water usually means schools must close and children are sent home. Teaching and learning are disrupted, leading to poorer educational achievement and outcomes (Harbison and Eric Hanushek 1992; White 2004). It is therefore important that schools that have unreliable water supplies or water shortages investigate how water supplies can be made more reliable, sufficient, and resilient.

One of the most effective ways of developing a more resilient water supply is through onsite water storage and rainwater harvesting (Thuy et al. 2019). Onsite water storage enables schools to operate when there are outages from municipal supplies. This storage can be fed from municipal systems, rainwater harvesting, or both.

DOI: 10.1201/9781032638102-12

As rainwater is effectively free, schools can benefit from reduced costs. Rainwater harvesting systems reduce the reliance on external water supplies and enable schools to use their supply to supplement or meet their requirements. In areas with reliable local water supply systems, rainwater harvesting systems may provide a supplementary or backup system, whereas in drought-stricken areas, rainwater harvesting systems may be the sole supply of water (Cook et al. 2013).

Internationally, there is an increasing interest in rainwater harvesting as a way of building more resilient water systems in cities and buildings. Rainwater harvesting systems can help address water shortages whilst avoiding the need to develop new large-scale water infrastructure. Ghaffarian Hoseini et al. (2016) point out that rainwater harvesting systems may be able to meet 80%–90% of household water consumption globally. Steffen et al (2012) show how the widespread installation of rainwater harvesting systems in cities can be used to avoid the need to construct new water supply systems. However, despite the potential of rainwater harvesting systems, they have not been widely implemented.

Given the benefits of rainwater harvesting systems and the disruption to education caused by water shortages, it is surprising that these systems are not more widely used in schools. The slow adoption of rainwater harvesting systems may be due to the following factors: First, rainwater harvesting may be considered expensive (Campisano et al. 2017; Akuffobea-Essilfie et al. 2020). Second, there are widely held concerns about the quality of water from rainwater harvesting systems (Rahman et al. 2014; Fewtrell and Kay 2007). Third, a lack of awareness and information may hinder the adoption of rainwater harvesting systems as a means of tackling local water shortages (Akuffobea-Essilfie et al. 2020; Sheikh 2020). Fourth, concerns about regulations and maintenance can also be an obstacle to implementation (Campisano et al. 2017).

However, once rainwater harvesting systems are installed, they appear to be well-liked and supported by users (Gould 1997; Fuentes-Galván et al. 2018; Sunkemo and Essa 2022). Regulations and incentives supportive of rainwater harvesting have also been found to be valuable in encouraging greater adoption of the technology (Domènech and Saurí 2011). In addition to the supply of water, rainwater harvesting systems can provide other benefits, such as being a useful teaching and learning resource in schools (Kerlin et al. 2015; Campisano et al. 2017).

This chapter aims to contribute to improved knowledge about rainwater harvesting systems in schools in water-scarce areas. It addresses misconceptions about rainwater harvesting by showing how rainwater harvesting systems work and can contribute to water resilience in schools. A case study is undertaken to investigate the design of rainwater harvesting systems in schools in water-scarce areas. The case study is used to address questions about the sufficiency and affordability of rainwater harvesting systems. Analysis of the results from the case study shows that rainwater harvesting systems can be developed that are affordable and provide sufficient water to meet a school's needs. Findings from the study are important as they indicate that implementing rainwater harvesting systems could help avoid disruption to education from water shortages and therefore make a valuable contribution to sustainable development in water-scarce countries.

8.2   SCHOOL RAINWATER HARVESTING

In most respects, rainwater harvesting systems in schools are similar to systems in commercial and domestic buildings; however, they are different in the following ways. Schools tend to be located on larger sites, which have a range of buildings with roof surfaces that can be used for rainwater collection. They also usually have hard surfaces, such as basketball courts, that can also be used for rainwater collection. They tend to have larger sites with available space that can be used for rainwater tanks. Schools also have varied populations over a year, as the school is occupied during term time and then empty during holiday periods. Within term times, occupancy can also vary widely, and numbers at a school may double, for instance, during a large sporting event. These factors are considered in the different elements of a school rainwater harvesting system that are described next.

The elements of a school rainwater harvesting system are shown in Figure 8.1. These are as follows:

A. Rainwater collection surface
B. Filtration system
C. Rainwater tanks
D. Additional filtration (if necessary)
E. Water uses include drinking, washing, cleaning, flushing toilets and irrigation

8.3   RAINWATER COLLECTION SURFACES

Schools are usually good candidates for rainwater harvesting because of the large roof areas that can be used as collection areas to harvest rainwater. They also usually have sufficient additional spaces where rainwater harvesting tanks can be located. In addition, schools have significant water requirements associated with cleaning, irrigation and flushing toilets, which can be readily met through rainwater harvesting.

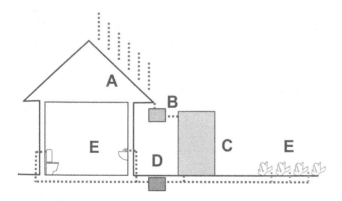

FIGURE 8.1   A school rainwater harvesting system. (Diagram by the author.)

TABLE 8.1

Runoff Coefficients (Farreny et al. 2011; Goel 2011)

Roof Type/Surface	Runoff Coefficient
Sloping corrugated metal roof sheeting and tiled roofing	0.9
Flat concrete roofing with gravel topping	0.8
Level cement surfaces, such as driveways and tennis courts	0.8
Pavements and roads	0.70–0.95
Parks and pastures	0.05–0.30

The main collection surfaces in schools used for harvesting rain are the roofs of buildings. However, schools may also have hard external surfaces such as external assembly areas, recreational yards and tennis, basketball and netball courts that can be used. Roofs work for rainwater harvesting as water can be gravity-fed into tanks located below them. Water harvested from yards and sports courts at ground level must be directed to sumps and underground tanks or pumped up to tanks. In general, hard surfaces at ground level may receive more dust and detritus which makes water from these surfaces more difficult to clean.

The quality and quantity of rainwater harvested off surfaces are dependent on the material of the collection surface and the extent to which it is exposed to dust and detritus that may affect the quality of runoff. The amount of runoff is dependent on the runoff coefficient of the surface. This is shown in Table 8.1 and represents the proportion of water harvested from a surface. Thus, from Table 8.1, one would expect to harvest 90% of the water from a corrugated roof and between 70% and 95% from a pavement or a road.

8.4 FILTRATION SYSTEM

To avoid dust and detritus from polluting rainwater harvesting, filters are normally included between the collection surface and rainwater tanks. These include 'first-flush' filters designed to remove dust and detritus that flow into the system with the first rain. To ensure that these filters are effective, they must be well-designed and regularly maintained. Therefore, schools should ensure these are regularly inspected and cleaned as necessary. In general, losses from filters are low unless they are poorly maintained and blocked. First flush filters can result in no flows to rainwater tanks if rainfall is very light, as initial water flows are diverted to remove dirt on collection surfaces.

8.5 RAINWATER TANKS

Once filtered, rain is stored in rainwater tanks. In the simplest systems, these tanks are placed directly under gutters and downpipes and are associated with the collection surface of a particular building. This arrangement is easy to construct and maintain. More complex systems direct water to large central tanks but are more difficult to construct and manage. Rainwater tanks are usually the most expensive element of a rainwater harvesting system, so care should be taken that these are correctly specified and sized.

Simple rainwater harvesting tanks are usually made of plastic and steel, are located on a concrete base and vary from 2,000 to 20,000 L. Large rainwater tanks may be constructed of concrete, be centrally located, be underground and fed from multiple collection surfaces and be over 20,000 L. In schools, it is important that tanks cannot be accessed by students but are designed to be easily inspected and cared for by maintenance staff. Rainwater tanks should include an indicator system that enables water volumes to be readily ascertained and managed.

The sizing of rainwater tanks is complex, and care should be taken not to rely on simple calculations, which may oversize rainwater tanks. Instead, modelling flows into rainwater harvesting systems and flows out through use over a year should be used to calculate the size of a tank. Examples of this are shown in the case study presented later in the chapter.

8.6 FILTERS

Where the quality of the water is important, for instance, for drinking, an additional filter may be included between the tank and the use of the water. If water will be used for drinking, it mustn't be polluted or dirty. This is achieved by ensuring that collection surfaces are clean and that any detritus or dirt is removed from runoff. The quality of water and its suitability for drinking can also be confirmed through tests. Water labs can test water for contaminants such as bacteria and heavy metals to ensure that it is safe for drinking.

8.7 WATER USES

Rainwater in schools can be used for cleaning, such as washing cooking utensils and equipment and mopping floors. It may also be used for flushing toilets and urinals, as well as for irrigation of sports fields, food gardens and ornamental beds. Rainwater can also be used for drinking if it is of a suitable quality (see Filters above). Where harvested rainwater is considered not suitable for drinking, this should be clearly labelled as "Non-potable water – not suitable for drinking".

Table 8.2 shows the volumes of water used for different activities in schools. Levels of consumption depend on the efficiency of the equipment. For instance, inefficient toilets and urinals can lead to levels of water consumption of over 30 L per occupant, while this can be less than 15 L in a school with efficient systems (Gibberd 2021).

Table 8.2 also shows that water use in schools can vary widely. Where water is extremely scarce, water use may only be 5–10 L per person. This is achieved by using water for drinking and some cleaning, with all other uses of water, such as toilet flushing, being avoided. In schools with a plentiful supply of water, water use may be over 70 L per person per day, as water is not only used for drinking and cleaning, but also for flushing toilets, irrigation and topping up swimming pools.

Water consumption figures of between 10 and 30 L per student per day are common in schools in water-scarce countries. For instance, Nunes et al (2019) indicate

TABLE 8.2
Water Uses in Schools (Gibberd 2021)

Water Use per Occupant per Occupied School Day	Litres
Drinking water[a]	2–3
Cleaning and washing[a]	2–6
Flushing toilets and urinals[b]	15–30
Irrigation, topping up swimming pools, and other uses[b]	10–30
Overall (Litres per student per day)	**10–70**

[a] Applicable to all schools.
[b] Not applicable to all schools, as some school may not have water-based sanitation, irrigation and swimming pools.

that water consumption in schools in Brazil ranges between 11 and 18 L per student per day. Figures are similar in Spain, where figures are between 4 and 16 L per student per day (Morote et al. 2020). In Italy, water use per student per day is between 10 and 70 L, with lower figures in primary schools and higher figures in secondary schools where there may be irrigated sports fields (Farina et al. 2011).

8.8 DESIGNING SCHOOL RAINWATER HARVESTING SYSTEMS

The first step in designing a rainwater harvesting system is to ensure that water is used efficiently in the school, as this helps to reduce the size of the rainwater harvesting system required. A review of Table 8.2 indicates that savings are unlikely to be achieved through reducing drinking water and cleaning water consumption. However, considerable savings can be achieved through more efficient fittings. For instance, water consumption for toilets can be cut in half by replacing inefficient 9-L WC flush cisterns with efficient 4.5-L flush cisterns and 2-L urinal flush mechanisms with 1-L flush mechanisms.

Even greater savings can be achieved by using very water-efficient sanitation systems such as aqua-privy and dry or composting toilet systems. Specifying these systems can reduce water consumption at schools by over 50%. Similarly, irrigation and the topping-up of swimming pools and ponds can be reduced. Where necessary, water use can be minimised through xeriscape landscaping (an approach that eliminates the need for irrigation) and by removing swimming and other pools.

8.9 RAINFALL PATTERNS

Local rainfall patterns are an important input for the design of a rainwater harvesting system. Rainfall data can be obtained at a range of frequencies, including hourly, daily, and monthly rainfall. Monthly rainfall data is the most readily available and is shown as mm per month, as indicated in Table 8.3 (Climate Data 2023).

A review of monthly data can be used to identify the type of rainwater harvesting system that will be required. Where rainfall varies widely over the year, with several

TABLE 8.3
Rainfall and Occupancy Days over a Year
(Climate Data 2023)

Month	Rainfall (mm/month)	Occupancy Days
Jan	49	10
Feb	53	20
Mar	56	23
Apr	47	13
May	28	23
Jun	27	17
Jul	28	10
Aug	43	23
Sep	37	20
Oct	53	16
Nov	61	22
Dec	51	9
Totals	**533**	**206**

months of limited rainfall, larger rainwater tanks and collection surfaces will be required to meet shortfalls during the dry season. Where rainfall is constant throughout the year, a much smaller system will suffice.

8.10   OCCUPANCY

The design of rainwater harvesting systems can be optimised by understanding patterns of water use in schools. This includes identifying the occupants of the school and when they are at the school.

School occupants include students, teachers, and administrative and maintenance staff. They may also include temporary occupants, such as parents who come for an evening event or a team from another school who may come for a sporting event. School terms usually mean that there are prolonged periods when the school may have very few occupants (i.e., during holidays).

Thus, while a residential building may be occupied for 365 days of the year, schools may only be occupied for 206 days of the year (the school terms), as shown in Table 8.3. Understanding these patterns of occupancy is valuable for ensuring systems are correctly sized. School rainwater harvesting systems may be overspecified if they are sized for full occupancy throughout the year and therefore will be unnecessarily expensive.

Ideally, patterns of occupancy of the building should align with rainfall patterns to ensure that water is available for use when people are at the school rather than having to provide extensive additional storage of water to cater for occupancy during dry periods. When this alignment happens, rainwater harvesting systems can be much smaller and, therefore, more affordable. Thus, where school calendars can be flexible, aligning occupancies with rainfall patterns could be used to develop more affordable and resilient rainwater harvesting systems.

8.11   CASE STUDY

A case study of a rainwater harvesting system is undertaken to investigate the affordability and sufficiency of these systems for meeting a school's water needs in a water-scarce area. The case study is near Loerie in the Eastern Cape, in South Africa. South Africa is classified as having an "extremely high" baseline water stress (Kuzma et al. 2023). Baseline water stress measures the ratio of total water demand to available renewable water supplies. According to the Koppen Geiger climate classification, the case study site in the Eastern Cape is in an area classified as BsH, Arid, Stepp and Hot Arid (CSIR 2023a).

Average monthly temperatures vary from 21.7°C in February to 14.2°C in July. The site currently has between 30 and 45 very hot days per year, with hot days being when maximum temperatures exceed 35°C (CSIR 2023b). As a result of climate change, average temperatures are projected to increase by 1°C–2°C by 2050. In addition, up to 10 additional very hot days relative to a baseline period of 1961–1990 are likely to be experienced (CSIR 2023b).

The annual average rainfall is 533 mm, and the driest month is June, with 27 mm. November has the highest rainfall, with an average of 61 mm (Table 8.3). Projections indicate that climate change will result in drier conditions and that a drop in annual rainfall of between 25 and 100 mm by 2050 is expected relative to a baseline period of 1961–1990 (CSIR 2023b). Over the last 8 years, the school site has experienced drought conditions, and water restrictions have been in place (Plaatjies 2023).

The case study school is shown in Figure 8.2. This shows large roof and yard areas available as collection surfaces. Figure 8.3 shows a plan of the school with the roof collection surfaces shown in dark grey. The roofs of the building are made of corrugated iron and have a runoff coefficient of 0.9 (Table 8.1). The roof collection surface area available for rainwater harvesting is 2,160 m^2 on a school site of 67,500 m^2. The school has a total of 210 occupants, with 202 being students and 8 full-time staff.

8.12   METHODS AND MATERIALS

To develop and analyse an off-grid resilient water system that could supply sufficient water for a school in a water-scarce area, the following steps were taken:

First, a brief for the water system is developed. The water restrictions of the last 8 years and worsening conditions projected in the future (presented above) for the site indicate that the key objective for the design of the water system is to achieve an off-grid resilient water supply that provides sufficient water to enable the school to operate despite local shortages.

Second, a suitable rainwater harvesting modelling tool that could simulate water consumption and rainwater harvesting water volumes over a year was selected. The Rainwater Use Model (RUM), shown in Figure 8.4, met this specification and was chosen. The RUM was developed by Gibberd (2020) and has been applied to a range of different building types.

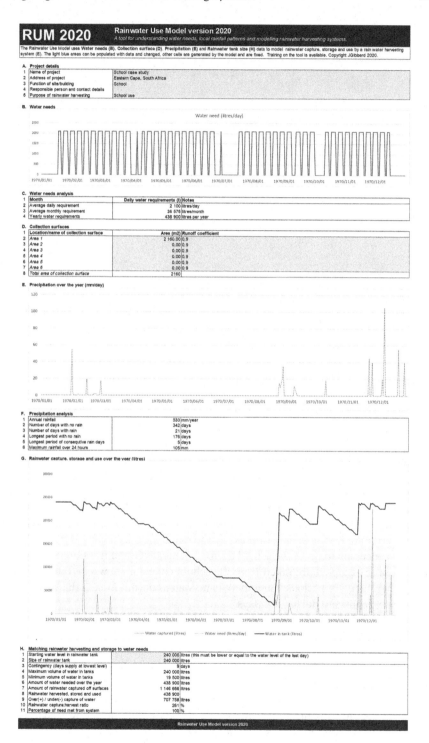

FIGURE 8.2 Photographs of the case study school. (Photographs by the author.)

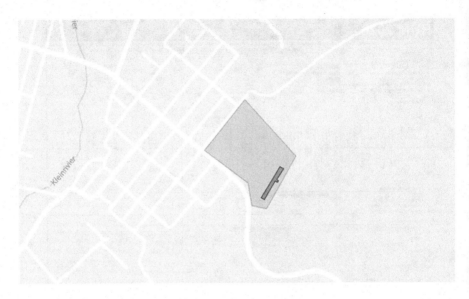

FIGURE 8.3 A plan of the school indicating the site and the roof collection surface. (Diagram by the author.)

Third, as the RUM uses daily rainfall patterns, appropriate local climate data were identified. Daily annual rainfall patterns were reviewed to identify a year when annual rainfall was similar to the average rainfall of 533 mm (Table 8.3). The year selected had an annual rainfall of 531 mm, which is entered into the RUM and shown in Figure 8.4, Graph E.

Fourth, levels of water consumption are established for the school. To ensure that the system would be resilient, it was decided to minimise water consumption at the school, and a volume of 10 L per person was allowed. This covers drinking water, basic hygiene practices and some cleaning and washing and is based on the figures in Table 8.2. Since it does not allow for flushing toilets, dry sanitation is required. The level of consumption is then multiplied by the number of occupants and the days the school is occupied (days of the school term) to generate Graph B, in Figure 8.4. Days the school is occupied are taken from the school calendar shown in Table 8.3. This shows that there are periods when there is no water consumption at the school. This is over weekends and during holidays.

Fifth, rainfall is multiplied by the area of the collection surface (2,160 m²) and by the runoff coefficient (0.9) to generate rainwater harvesting volumes per day over the year. This is shown in Graph G in Figure 8.4.

Sixth, water tank sizes are entered into the RUM on a trial-and-error basis until the required performance is achieved (in this case, off-grid performance). Figure 8.4 shows that a volume of 240,000 L achieves sufficient storage for the rainwater harvesting system to fully meet the water needs of the school. The yellow line in Graph G in Figure 8.4 shows rainwater volumes in the tanks over a year.

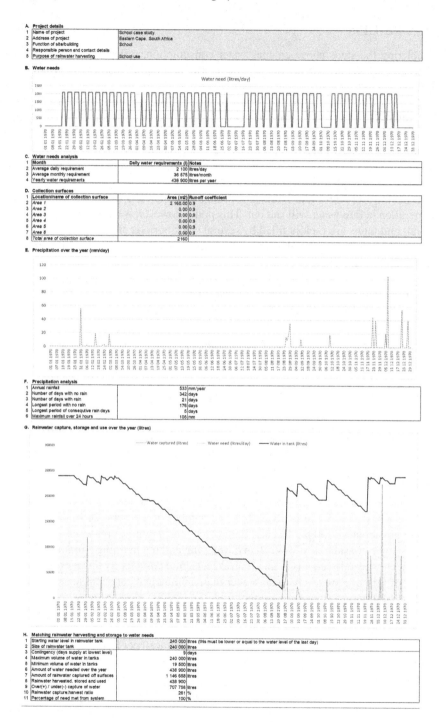

FIGURE 8.4 Rainwater use model (RUM) results for the case study school. (Reports by the author)

8.13   RESULTS

Completing the steps above enables water consumption, rainwater harvesting, and rainwater tank volumes to be simulated over a year, and this is shown in Graph G of Figure 8.4.

Graph G shows how, at the beginning of the year, the rainwater tanks are full and hold 240,000 L (the yellow line). However, ongoing consumption (the orange line) and lack of rainfall (the blue line) mean that volumes drop steadily. There is some respite in July when water is not used as the school is on holiday, and therefore rainwater tank volumes remain steady. However, after this, volumes continue to drop and only recover in September, when there is rainfall.

At the lowest point, water levels in the tanks are only sufficient for a further 9 days of operation before water will run out. This point, however, is not reached, as there is rainfall and tanks are rapidly filled by several short periods of heavy rainfall.

The results prompt several additional questions about the design of the rainwater harvesting system. These are:

- How much would the rainwater harvesting system cost?
- How could you increase the contingency or safety buffer of the system to help ensure that water would not run out in a very dry year?
- If there was a decision to have water-borne sanitation, what would this mean for the design of the system?

8.14   COSTS OF A RAINWATER HARVESTING SYSTEM

The design of the rainwater harvesting system is very simple and consists of roofs, gutters and rainwater tanks underneath the gutters, with some plumbing from tanks to locations where water can be used for drinking and washing. Therefore, the capital costs of the system consist of plumbing from gutters to tanks and from tanks to where water is used, stand-alone 10,000-L plastic tanks and concrete bases for the tanks. The costs of this infrastructure are estimated to be about $20,000, as shown in Table 8.4. If this figure is divided by the number of learners (200), the costs per learner are about $1,000 per learner. In new schools in remote areas, this capital cost is likely to be considerably lower than developing and maintaining a mains grid connection. However, in schools in an urban area, capital costs for a rainwater harvesting system of this size are likely to be higher than a municipal connection, as only short links to existing infrastructure are required.

An analysis of the operating costs and payback period of the rainwater harvesting system is also undertaken. Table 8.4 shows that the cost of water from the local authority is R15 per kilolitre. This increases to about R20 per kilolitre under drought conditions. These tariffs can be multiplied by the annual consumption of water to get annual water costs for the school of between R6,600 and R8,600/year. Given that the capital cost of the rainwater harvesting system is R490,000, the payback period is between 50 and 70 years, as shown in Table 8.4. The very long payback period indicates that the rainwater harvesting system does not make sense if viewed purely from a commercial perspective. However, viewed from an education perspective,

TABLE 8.4

Capital Costs and Payback Periods of Rainwater Harvesting System at a School (Gibberd 2023; Kouga 2023)

Component	Cost (Rand)	Number	Totals
Tanks	15,000	24	360,000
Bases	3,000	24	72,000
Collection plumbing	1,000	24	24,000
Distribution plumbing	1,000	26	24,000
Drinking water filtration	2,000	6	12,000
Capital cost in South African Rand (R)			492,000
Capital costs in USD, at an exchange rate of R17 to the USD			20,118
Water tariff schools - normal (R/KL)			15.00
Water tariff for schools - drought (R/KL)			20.12
Water consumption per year (L/year)			440,000
Payback for normal conditions (years)			74
Payback for drought conditions (years)			57

costs may be justified in relation to the negative long-term and large-scale impacts associated with disruption and poor education outcomes that are avoided.

8.15   DETERMINING AND INCREASING SAFETY MARGINS OF THE RAINWATER HARVESTING SYSTEM

As shown in the introduction, there are severe negative impacts if a school runs out of water. So, what can be done to avoid this? This is addressed in the case study by ensuring that simulations are based on data from an average rainfall year and through the concept of contingencies.

The contingency level in the RUM is shown in Table H of Figure 8.4 and is the number of days that operations could continue at the school when water levels in rainwater tanks are at their lowest levels. In the current design, the contingency is 9 days. This means that without the rainfall in September, the water at the school would run out.

The contingency, or margin of safety, for the rainwater harvesting system, can be increased in this case by changing the capacity of the rainwater harvesting system or reducing consumption levels. Adding rainwater harvesting capacity to a volume of 280,000 L in the system increases the contingency to 28 days, for the same climatic year.

8.16   THE IMPLICATIONS OF WATER-BORNE SANITATION FOR RAINWATER HARVESTING SYSTEMS

The implications of water-borne sanitation for the case study are modelled by increasing the levels of water consumption from 10 to 25 L per person per day (based on Table 8.2). This dramatically increases water consumption at the school and the

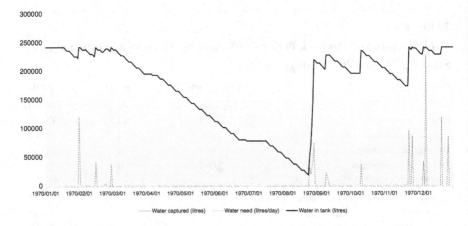

FIGURE 8.5 Rainwater use model (RUM) results for 5,000 m² collection surfaces and 600,000-L rainwater tanks. (Report by the author)

rainwater harvesting system must be redesigned to be able to meet this demand. Achieving this demand requires a much larger collection surface and rainwater harvesting tanks. Figure 8.5 shows a solution based on a 5,000 m² collection surface and water storage of 600,000 L. This system provides a similar contingency to the original design of about 9 days. The example shows that it is possible to design a rainwater harvesting system for the school that accommodates waterborne sanitation. However, this system requires the use of external hard paving, sumps, pumps, and rainwater tanks that are over double the capacity of the system required for dry sanitation. This will result in a far more expensive system.

8.17 DISCUSSION

An analysis of the case study provides the following findings for the design of rainwater harvesting systems in schools: First, the case study shows that it is possible to design rainwater harvesting for a school in an area with only about 500 mm of annual rainfall that enables a school to be off-grid and not reliant on external water sources. This system allows schools to continue to operate even if there are local water shortages.

Second, it shows that the cost of the system is approximately $1,000 per student. By being off grid, the school could achieve operational savings by not having to pay for connection charges and water consumption from municipal supplies. Water charges can be increased by 30% during drought periods (Kouga 2023). While the system has long payback periods (see Table 8.4), it does enable the school to have control over water costs and to budget and manage them more easily.

Third, the case study shows water consumption must be reduced to about 10 L per person per day in order for a simple rainwater harvesting system to cater for all of the school's water needs. This can only be achieved by avoiding waterborne sanitation, and therefore the school would have to have a dry sanitation system such as an aqua privy or composting system. This may be a useful teaching tool, as it can be used to show how compost or fertiliser from the system can be used to maintain fertility in soils. This is particularly relevant to schools in farming areas.

Fourth, the case study shows the implications of waterborne sanitation. To accommodate waterborne sanitation, a figure of 25 L per person per day is used, and therefore, much greater volumes of water are consumed. To cater for this, a much larger rainwater harvesting system is required, and the case study shows that the area of rainwater collection has to more than double to 5,000 m². Similarly, rainwater harvesting tanks must increase to 600,000 L. Thus, the case study shows that while it is possible to accommodate water-borne sanitation with a rainwater harvesting system, it requires a much bigger and more complex system. For instance, additional collection surfaces in the form of external hard surfaces such as netball courts and yards would be required. These, in turn, require additional filtration and pumps, which would need to be supplied with electricity and maintained. This level of increased cost and complexity may be difficult to justify in schools with limited resources and capacity for maintenance.

Fifth, the case study shows how the risk of running out of water can be addressed by using average rainfall years and contingency targets. It shows how the rainwater harvesting system can be modified to improve contingency periods. This aspect enables a school to understand the risks of the system and plan for them. Thus, to make the system more affordable, a school may have shorter contingency periods in the initial system. However, over time, contingencies could be increased and the risk of running out of water reduced by adding more storage and catchment areas to the system. In this way, the school can increase the resilience of the system to enable it to cope with climate change.

8.18 CONCLUSIONS AND RECOMMENDATIONS

The study shows that simple rainwater harvesting systems can be designed to meet all the water needs of a school in a water-scarce climate. This requires water consumption levels at schools to be around 10 L per person per day and dry sanitation to be used. The study shows that the cost of this system, at about $1,000 per student/person, is not prohibitive and could make a valuable contribution to local sustainable development by avoiding disruption and improving education outcomes.

It confirms that waterborne sanitation places large, increased demands on water, which can be met by a rainwater harvesting system. While this rainwater harvesting solution is possible, the study suggests that the additional costs and complexity associated with achieving it may not make it worthwhile. Instead, lower rates of water consumption and simpler rainwater harvesting systems that can be expanded easily, combined with dry sanitation systems, are recommended.

A significant finding of the study is the potential for schools in water-scarce areas to develop water-resilient off-grid water systems at low cost by using rainwater harvesting. Implementing rainwater harvesting could significantly reduce disruption to education from water shortages and help reduce the pressure on main water systems.

A recommendation from the study is that further research on rainwater harvesting systems in schools in water-scarce areas be undertaken to develop effective guidance that will enable schools to improve the resilience of their water supplies and ensure that disruptions to education because of water shortages are avoided.

ACKNOWLEDGEMENTS

Valuable comments and inputs by the editors that have improved the chapter are acknowledged.

REFERENCES

Akuffobea-Essilfie, M., Williams, P.A., Asare, R., Damman, S. and Essegbey, G.O. 2020. Promoting rainwater harvesting for improving water security: Analysis of drivers and barriers in Ghana. *African Journal of Science, Technology, Innovation and Development*, 12(4): 443–451.

Campisano, A., Butler, D., Ward, S., Burns, M.J., Friedler, E., DeBusk, K., Fisher-Jeffes, L.N., Ghisi, E., Rahman, A., Furumai, H. and Han, M. 2017. Urban rainwater harvesting systems: Research, implementation and future perspectives. *Water Research*, 115: 195–209.

Climate Data. 2023. Loerie climate: Weather Loerie & temperature by month. https://en.climate-data.org/africa/south-africa/eastern-cape/loerie-919707/ (accessed December 12, 2023).

Cook, S., Sharma, A., Chong, M. 2013. Performance analysis of a communal residential rainwater system of potable supply: A case study in Brisbane, Australia. *Water Resources Management*, 27: 4865–4875

CSIR. 2023a. Köppen-Geiger zones. https://csir.maps.arcgis.com/apps/webappviewer/index.html?id=22cb03d8bd244f6ea3dd67e946f0ce1e (accessed December 21, 2023).

CSIR. 2023b. GreenBook risk profile tool. CSIR: Pretoria. https://riskprofiles.greenbook.co.za/ (accessed December 12, 2023).

Diedhiou, A., Bichet, A., Wartenburger, R., Seneviratne, S.I., Rowell, D.P., Sylla, M.B., Diallo, I., Todzo, S., N''datchoh, E.T., Camara, M. and Ngatchah, B.N. 2018. Changes in climate extremes over West and Central Africa at 1.5°C and 2°C global warming. *Environmental Research Letters*, 13(6):065020.

Domènech, L. and Saurí, D. 2011. A comparative appraisal of the use of rainwater harvesting in single and multi-family buildings of the Metropolitan Area of Barcelona (Spain): Social experience, drinking water savings and economic costs. *Journal of Cleaner Production*, 19(6–Farina, M., Maglionico, M., Pollastri, M., & Stojkov, I, 2011. Water consumptions in public schools. Procedia Engineering, 21, 929-938: 598–608.

Farreny, R., Morales-Pinzón, T., Guisasola, A., Tayà, C., Rieradevall, J. and Gabarrell, X. 2011. Roof selection for rainwater harvesting: Quantity and quality assessments in Spain. *Water Research*, 45(10):3245–3254.

Fewtrell, L. and Kay, D. 2007. Microbial quality of rainwater supplies in developed countries: A review. *Urban Water Journal*, 4(4):253–260.

Fuentes-Galván, M.L., Ortiz Medel, J. and Arias Hernández, L.A. 2018. Roof rainwater harvesting in Central Mexico: Uses, benefits, and factors of adoption. *Water*, 10(2): 116.

Goel, M.K. 2011. Runoff coefficient. In: Singh V. P., Singh P., Haritashya U. K. (eds) *Encyclopedia of snow, ice and glaciers. Encyclopedia of earth sciences series.* Springer, Dordrecht.

Gibberd, J. 2020. An alternative rainwater harvesting system design methodology. In: *The sustainability handbook*, Volume 1. Alive 2 Green Publishers, Cape Town, South Africa.

Gibberd, J. 2021. Rapid identification and evaluation of interventions for improved water performance at South Africa schools. *Sustainability in Energy and Buildings*, 2020: 173–182.

Gibberd, J. 2021. Rapid Identification and Evaluation of Interventions for Improved Water Performance at South Africa Schools. In: Littlewood, J., Howlett, R.J., Jain, L.C. (eds) *Sustainability in Energy and Buildings 2020. Smart Innovation, Systems and Technologies*, vol 203. Springer, Singapore. https://doi.org/10.1007/978-981-15-8783-2_14

Gibberd, J. 2023. Costings based on personal communication with a Quantity Surveyor.

Gould, J. 1997. Catching up--upgrading Botswana's rainwater catchment systems. *Waterlines*, 15(3):13–14.

Harbison, R. and Eric Hanushek, E. 1992. *Educational performance of the poor: Lessons from rural Northeast Brazil*. Oxford University Press for the World Bank.

Hoseini, A.G., Tookey, J., Yusoff, S.M. Hassan, N.B. 2016. State of the art of rainwater harvesting systems towards promoting green built environments: A review. *Desalination and Water Treatment*, 57 (1): 95–104

IPCC. 2022. Summary for policymakers climate change 2022: Impacts, adaptation and vulnerability. *Contribution of Working Group II to the Sixth Assessment Report of the Intergovernmental Panel on Climate Change*. 2022; Cambridge University Press, Cambridge, UK and New York, NY: 3–33.

Jasper, C., Le, T.T. and Bartram, J. 2012. Water and sanitation in schools: A systematic review of the health and educational outcomes. *International Journal of Environmental Research and Public Health*, 9(8): 2772–2787.

Kerlin, S., Santos, R. and Bennett, W. 2015. Green schools as learning laboratories? Teachers' perceptions of their first year in a new green middle school. *Journal of Sustainability Education*, 8: 1–11.

Kouga. 2023. Tariffs 2022/2023. https://www.kouga.gov.za/download/4977 (accessed December 21, 2023).

Kuzma, S., Bierkens, M.F., Lakshman, S., Luo, T., Saccoccia, L., Sutanudjaja, E.H. and Van Beek, R. 2023. Aqueduct 4.0: Updated decision-relevant global water risk indicators. Wri.org, (2023). https://www.wri.org/research/aqueduct-40-updated-decision-relevant-global-water-risk-indicators (accessed December 21, 2023).

Makki, A.A., Stewart, R.A., Beal, C.D. and Panuwatwanich, K. 2015. Novel bottom-up urban water demand forecasting model: Revealing the determinants, drivers and predictors of residential indoor end-use consumption. *Resources, Conservation and Recycling*, 95:15–37.

Morote, Á.F., Hernández, M., Olcina, J. and Rico, A.M. 2020. Water consumption and management in schools in the City of Alicante (Southern Spain) (2000–2017): Free water helps promote saving water? *Water*, 12(4):1052.

Nunes, L.G.C.F., Soares, A.E.P., Soares, W.D.A. and da Silva, S.R. 2019. Water consumption in public schools: A case study. *Journal of Water, Sanitation and Hygiene for Development*, 9(1):119–128.

Plaatjies, R. 2023. Water restrictions on Kouga-Loerie sub-system lifted. https://www.news24.com/news24/community-newspaper/kouga-express/water-restrictions-on-kouga-loerie-sub-system-lifted-20231011 (accessed December 21, 2023).

Sheikh, V. 2020. Perception of domestic rainwater harvesting by Iranian citizens. *Sustainable Cities and Society*, 60; 102278.

Steffen, J., Jensen, M., Pomeroy, C.A., Burian, S.J. 2012. Water supply and stormwater management benefits of residential rainwater harvesting in U.S. cities. *Journal of the American Water Resources Association*, 49 (4): 810

Sunkemo, A. and Essa, M. 2022. Exploring factors that affect adoption of storage-based rainwater harvesting technologies: The case of Silte Zone, Southern Ethiopia. *Proceedings of the International Academy of Ecology and Environmental Sciences*, 12(3): 144–156.

Thuy, B.T., Dao, A.D., Han, M., Nguyen, D.C., Nguyen, V.A., Park, H., Luan, P.D.M.H., Duyen, N.T.T. and Nguyen, H.Q. 2019. Rainwater for drinking in Vietnam: Barriers and strategies. *Journal of Water Supply: Research and Technology-AQUA*, 68(7): 585–594.

UN-Habitat and IHS-Erasmus University Rotterdam. 2018. The State of African Cities.

UNESCO World Water Assessment Programme. 2019. The United NationsWorld Water Development Report 2019: Leaving No One Behind. Paris, UNESCO.

Watkins, K., 2006. Human Development Report 2006-Beyond scarcity: Power, poverty and the global water crisis. UNDP Human Development Reports (2006).

Wensley, A. and Mackintosh, G. 2015. *Water risks in South Africa, with a particular focus on the "Business Health" of municipal water services.* DHI-SA 2015 Annual Conference.

White, H. 2004. *Books, buildings, and learning outcomes: An impact evaluation of World Bank support to basic education in Ghana.* OED World Bank.

World Health Organization. 2009. *Global health risks: Mortality and burden of disease attributable to selected major risks.* World Health Organization.

Case Study 3

Mapping of Suitable Green Water Harvesting Interventions at Arid Ecosystems to Enhance Water-food-ecosystem Nexus

A Case from Jordan

Lubna Mustafa AlMahasneh and Doaa Ismail Abuhamoor

SUMMARY

Water and food security are the key challenges of climate change. In Jordan, drought and flash flood risks are also influenced by climate change and other anthropological factors. Green Water Harvesting (GWH) can be a strategic and adaptive solution to climate change and can enhance resilience in arid and semi-arid ecosystems. GWH involves capturing and efficiently utilising blue rainwater runoff to sustainably support agro-pastoral systems and promote drought tolerance. Water harvesting is a proven technology that may be effectively used to increase soil moisture content, vegetation cover, productivity and groundwater recharge. GWH suitability mapping for the Jawa site in Jordan was developed using an innovative approach that integrates soil and climatic data. The focus of this case study was to identify suitable locations for different rainwater harvesting interventions using Geographic Information Systems (GIS). The main biophysical parameters of slope, soil depth, soil texture, and stoniness were used to assess the suitability for rainwater harvesting. Criteria for each intervention were applied, and suitability maps were produced for selected rainwater harvesting interventions in micro- and macro-catchment land systems. As a result, the land suitability evaluation classified land as suitable and not suitable, corresponding to different potentials for a particular intervention. The produced maps can be beneficial for land managers, decision makers and farmers, and they can provide planning guidance for sustainable GWH and for enhancing the water-food-ecosystem nexus in Jordan.

PROBLEM STATEMENT

The escalation of drought and flash flood risks due to climate change, compounded by various environmental factors, underscores the imperative need for implementing GWH. This strategy serves as a pivotal and adaptive measure for combating the impacts of climate change, particularly in arid and semi-arid ecosystems. GWH acts as a strategic solution, instrumental in optimising resilience by effectively managing water resources, fostering biodiversity, and sustaining ecosystem health. Its implementation is not merely advantageous; it is an essential tool to fortify these vulnerable ecosystems against the detrimental effects of an evolving climate.

GWH maps serve as crucial tools catering to a diverse range of users, from decision-makers to researchers and farmers. These maps act as comprehensive guides, enabling land users to optimise their land based on its specific suitability for various water harvesting techniques. They achieve this by integrating essential soil and climatic data. Moreover, these maps take into account the priorities of farmers, governmental bodies, and policymakers. According to statistics from the NARC, the platform hosting these water harvesting maps has been beneficial for more than 1,000 guest users. Additionally, within the past year, 1,500 users have actively engaged with and utilised these maps, underlining their practical significance and utilisation among stakeholders.

ACTIONS TAKEN

1. *Site selection and description:* The Jawa site in this case study, which is located in Mafraq governorate with coordinates of 32°20′06″N 37°00′12″, is the oldest proto-urban development in Jordan, dating from the late 4th millennium BC (Early Bronze Age). It is located in one of the driest areas of the Black Desert (Harrat al-Shamah) of Eastern Jordan and receives an annual rainfall of 150 mm. Remains of dams have been found at this site, the largest of which is a masonry gravity dam and the oldest known dam in the world, which was used as protection from flash floods.

2. *Watershed characterisation:* Once the case study site was selected, watershed characterisation was used to collect essential data for identifying potential sites within the case study area suitable for various water harvesting interventions. Data collection from different sources and formats were gathered as follows:
   – Existing soil data were used from the national soil map survey, which provided information in the watershed about the following parameters: texture, stoniness percentage, vegetation cover and soil depth.

  - Slope classes, stream networks and watershed boundaries were derived based on the Digital Elevation Model (DEM) of 30 m pixel resolution.
3. *Suitability analyses:* A suitability analysis model was applied for the selected water harvesting interventions: Contour Ridges and Vallerani for shrubs and Runoff Strips for field crops and Marabs (wadi beds). By determining the physical requirements for each type of water harvesting intervention, the water harvesting intervention requirements were matched with the land condition to identify potential areas suitable for that water harvesting intervention.

RESULTS

The Jawa case study represents the arid conditions of ecosystems. The main output of the characterisation process was a suitability map indicating the distribution of areas suitable for selected water harvesting techniques within the site. The biophysical soil parameters and slope classes were classified according to the requirements for each technique. Criteria for each parameter were applied, and suitability maps were produced for rainwater harvesting interventions in micro- and macro-catchment land systems. As a result, the land suitability evaluation classified land as suitable and not suitable, corresponding to a different potentials for a particular intervention of Contour Ridges and Vallerani for shrubs, Runoff Strips for field crops and Marabs (wadi beds) for trees and field crops. As such, it was possible to produce a road map of potential GWH sites for scientific research and investment in watershed management in Jawa.

SUSTAINABLE DEVELOPMENT GOALS AND IMPACT

This case study contributes to the following Sustainable Development Goals: SDG 15: Life on land, SDG 13: Climate action and SDG 2: Zero hunger. Implementing GWH techniques can have various positive impacts such as:

  - Increased Agricultural Production: By conserving and utilising rainwater more efficiently, farmers can enhance crop yields even with limited water resources.
  - Reduced Dependency on Irrigation: GWH reduces reliance on traditional irrigation methods, which helps conserve the limited freshwater sources.
  - Improved Soil Quality: Techniques like soil moisture conservation and agroforestry can enhance soil fertility and structure, making it more resilient to drought and improving long-term agricultural productivity.
  - Ecological Benefits: These interventions can contribute to ecosystem restoration, biodiversity conservation, and reduced soil erosion, thereby positively impacting the environment.
  - Community Empowerment: Implementing such interventions may involve local communities, empowering them through knowledge and sustainable practices, thus fostering resilience to water scarcity.

While GWH interventions offer promising solutions, their widespread adoption and impact can vary based on factors like geographical location, socioeconomic conditions, technological feasibility, and the level of community engagement. In Jordan, the successful implementation of GWH techniques could significantly alleviate pressure on water resources for both domestic and agricultural purposes, ultimately contributing to increased food security and ecological sustainability.

ADVICE FOR OTHERS LOOKING TO DO SOMETHING SIMILAR

Implementing water harvesting interventions is a valuable and sustainable strategy to combat water scarcity. To guide similar initiatives, consider the following:

WHAT TO DO

- Assess local conditions: Understand local climate, rainfall patterns, and topography.
- Identify potential water sources and high runoff areas.
- Choose appropriate techniques: Select methods like rainwater harvesting, check dams, contour bunding, and rooftop harvesting. Combine techniques for a comprehensive approach.
- Community involvement: Involve the local community from the beginning. Encourage participation in planning, implementation, and maintenance.
- Educate and Raise Awareness: Conduct awareness campaigns about water conservation and harvesting benefits.
- Provide training on maintenance and sustainable use: Integrate with land use planning.
- Incorporate water harvesting into regional land use planning: Collaborate with local authorities for policy integration.
- Regular maintenance: Establish a maintenance plan for longevity and efficiency. Train locals or hire teams for regular inspections and repairs.
- Monitor and evaluate: Implement a monitoring system to assess performance. Regularly evaluate the impact on water availability, crop yields, and community well-being.

WHAT TO AVOID

- One-size-fits-all approach: Tailor solutions to local conditions; avoid standardised approaches.
- Ignoring traditional knowledge: Incorporate local wisdom and traditional practices.
- Incomplete community engagement: Include the community in decision-making to prevent resistance.
- Neglecting maintenance plans: Establish a sustainable maintenance strategy to prevent deterioration.

- Ignoring environmental impact: Implement interventions with environmental sustainability in mind.
- Overlooking water quality: Ensure safe drinking water if using rainwater harvesting.

Following these guidelines will contribute to the success of water harvesting interventions that are able to promote sustainable water management and address water scarcity challenges in your region.

Case Study 4
Rainwater Harvesting Case Study from Gansu Province, China

John Gould

SUMMARY

This case study describes one of the world's largest and most successful rainwater harvesting projects, located in the semi-arid interior of central China. For centuries people living on the loess plateau in Gansu Province suffered from severe water shortages due to the lack of surface and ground water sources in this region. Traditional water sources were composed of water cellars excavated into the silty loess soil and annually lined with clay with volumes of up to 20 m³. Rainwater harvested from the ground and stored in these tanks was often insufficient to provide a reliable year-round water supply, especially during droughts. As a result, people had to spend large amounts of time and energy fetching water from distant sources. In the late 1980s the Gansu Research Institute for Water Conservancy (GRIWAC) was established and tasked with identifying possible solutions to addressing the water crisis. The goal of the programme was to research, field test, pilot and implement improved rainwater harvesting systems based on upgrading traditional technologies to provide safe, reliable, affordable and sustainable water sources. GRIWAC was successful in developing improved rainwater catchments and subsurface tank designs that communities could both afford and construct themselves. Due to support from the Gansu Provincial Government, which provided free construction materials and training to communities, and the active participation of beneficiaries who provided free labour, the new upgraded designs were widely adopted across the region, improving water supplies for millions of people (Figure C.S.4.1).

PROBLEM STATEMENT

Since the late 1980s, a major transformation has taken place in remote, water-stressed rural communities located on the Loess Plateau in China's arid interior, where mean annual rainfall ranges from 250 to 550 mm. This has been accomplished through the implementation of sustainable rainwater harvesting, low-cost greenhouses and

DOI: 10.1201/9781032638102-14

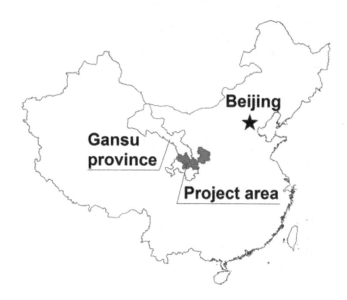

FIGURE C.S.4.1   Map showing location of Gansu Province, China and Project Area.

irrigation technologies. The result has been the widespread alleviation of poverty as well as improved diets, health, and livelihoods across the whole region.

The beneficiaries of the project include over 2 million people living in remote rural areas of Gansu province. Before the implementation of the programme, these were amongst the poorest communities in China. During a severe drought in 1995, many of these communities trekked up to 50 km to distant water sources or depended on emergency supplies delivered by water tankers. This prompted the government to support the programme to implement the widespread use of rainwater harvesting as the primary domestic water source in all rural water-stressed areas (Figure C.S.4.2).

ACTIONS TAKEN

At the heart of this programme was an approach which built on traditional knowledge of rainwater harvesting designs refined over centuries. By applying modern materials and research methods, these were improved to increase the quantity and reliability of household water supplies. Traditional *Shujiao* water cellars, which had been excavated in the loess soils and annually lined with clay, were upgraded using cement for both the subsurface tank and the catchment apron, or a tiled roof catchment. Upgrading the tank greatly reduced the annual maintenance requirements, and using an enlarged cement and/or tiled roof catchment with a high runoff coefficient greatly increased the efficiency of the rainwater harvesting system.

The Provincial Government provided 1.5–2 tons of cement per household, as well as training and other materials, to local communities, and the beneficiaries constructed the systems themselves. This allowed the construction of two 20-m^3 tanks for every participating family (Figure C.S.4.3).

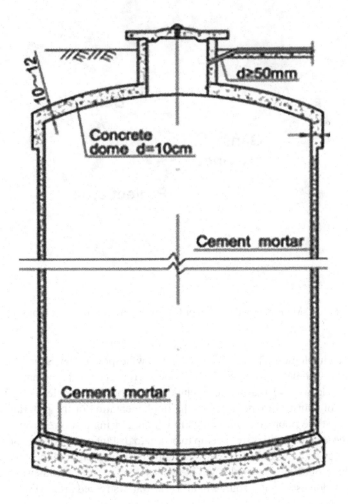

FIGURE C.S.4.2 Upgraded sub-surface water cellar design (Drawn by Qiang Zhu et al, GRIWAC, Lanzhou, China 2005).

OUTCOMES AND CHALLENGES

Overall, the rainwater harvesting programme in Gansu was successful in relieving the problem of domestic water shortages for the majority of the rural population on the Loess Plateau which had previously been subject to perennial water stress. Following the drought in 1995, additional resources from the Provincial Government led to a rapid expansion of the project. While data on the exact numbers of people who benefited from improved domestic rainwater supplies is limited, the Centre for Science and Environment website (CSE 2023) states that by 2000, a total of 2,183,000 rainwater tanks had been built with a total capacity of 73 million m³ in Gansu Province, supplying drinking water for 1.97 million people.

Among the challenges faced by the programme were that in order to meet target water supply and reliability levels of 20–40 L daily with 95% reliability, the combined

FIGURE C.S.4.3 Tiled roof and cement courtyard catchments (above) and a simple hand-pump for extracting water from the subsurface tank (below). (Photos by the author).

roof and courtyard catchment areas needed to range from 230 to 440 m^2 depending on local rainfall conditions. In many cases, the catchment areas were undersized, thereby reducing the amount of water and reliability of supply.

Since much of the rainwater was harvested from the ground, there were also some water quality issues, and while no serious health impacts were reported, national water quality limits for drinking water were sometimes exceeded.

SUSTAINABLE DEVELOPMENT GOALS, IMPACT AND REPLICATION

The programme has impacted several SDGs, but particularly addressed SDG 1: No Poverty, SDG 3: Good Health, SDG 6: Clean Water and SDG 11: Sustainable Communities.

The impact of the project on the beneficiaries in rural Gansu Province has been truly transformative, and chronic water shortages are now a thing of the past.

The provision of household domestic rainwater supplies, however, has been just one component of a broader initiative to also improve agriculture and livelihoods through the introduction of a range of appropriate technologies, including low-cost greenhouses and solar cookers (discussed below). The total combined effect of all these innovations has been a dramatic improvement in the lives of the beneficiaries.

The project in Gansu has now been fully implemented and provides an inspirational example of what can be achieved when government and communities work together. While the rainwater harvesting systems used in Gansu will not be appropriate in other parts of the world, many of the principles underpinning the success of this project can be applied elsewhere.

A great deal can be learned from the success of the Gansu case study, particularly by considering some of the key elements that underpinned the programme. These include:

- the technology was built on local knowledge and traditional wisdom accumulated from centuries of harvesting rainwater in the region.
- careful research, field testing and piloting done to ensure the viability of the new technology and its implementation.
- a bottom-up approach was supported by top-down support, with the local government providing cement and training, while the community was involved in planning, implementing, operating and maintaining the RWH systems.

While adopting similar universal principles for RWH projects elsewhere will undoubtedly go a long way towards ensuring their success, it is important to remember that trying to transfer technologies and implementation strategies between regions or countries is often fraught with difficulties. Invariably, there will always be a need to adapt technologies and methods of implementation to local conditions, since successful projects always depend as much on being socially, culturally, and economically appropriate as on their technical merits.

FURTHER BACKGROUND INFORMATION

The Gansu rainwater harvesting programme was conducted in concert with a number of other related initiatives that contributed to both the health and livelihoods of rural communities across the Loess Plateau. These included the development of low-cost solar cookers, which, while being produced commercially in local factories, were easily affordable to most households. These enabled water to be boiled and meals cooked at no cost, using free energy from the sun. This technology was highly suited to the generally treeless Loess Plateau region with its high annual sunshine hours. As most people consumed tea as their main beverage, boiling the water from the rainwater tanks also ensured it was safe to drink (Figure C.S.4.4).

Another related programme was the introduction of low-cost greenhouses constructed using mud walls, steel frames and plastic sheeting. These greenhouses, with areas of about 300 m^2, were oriented to make maximum use of passive solar warming to start growing crops early in the spring. Their plastic roofs and sometimes

FIGURE C.S.4.4   Low-cost solar cooker used for cooking and boiling water. (Photo by the author).

a surrounding cement apron were used as a catchment to harvest rain, which was stored in one or two 20- or 30-m³ subsurface tanks, similar to those being constructed at households for domestic water supply. The water in these tanks has been used to irrigate a variety of crops, including vegetables, fruit, flowers and mushrooms (Figure C.S.4.5).

These low-cost greenhouses have become widespread in communities across rural Gansu and have helped transform the rural economy. Often, two crops a year can be produced in them, and they have helped to improve the diets, livelihoods and health of the population.

In recent decades, several hundred million people have been lifted out of poverty in rural China. The low-cost and appropriate technologies outlined in this case study provide examples of some of the approaches involving the harvesting of rainwater and solar energy that have been used to achieve this impressive feat. The project clearly demonstrates the implementation of a model of sustainable development at

FIGURE C.S.4.5 Low-cost greenhouse with crops irrigated using water harvested from the roof and surrounding cement apron. (Photo by the author).

scale using affordable technologies. This is an approach that could potentially be adapted and replicated in other semi-arid regions around the world. For readers interested in a more detailed description of this case study, this can be found in the book: Every Last Drop (Zhu, Yuanhong and Gould 2012).

REFERENCES

CSE - Centre for Science and Environment (2023), Rainwater Harvesting, International Rural Case studies, China. https://www.rainwaterharvesting.org/international/china.htm
Zhu, Q., Yuanhong, L. and Gould, J. (2012). *Every Last Drop: Rainwater Harvesting and Sustainable Technologies in rural China*, Practical Action Publishing, U.K.

Case Study 5
Inter-seasonal Rainwater Harvesting Project at a School in Zimbabwe

Lynn McGoff, John A. J. Mullett,
Sandile Mtetwa, and Admire T. Baudi

SUMMARY

In Sub-Saharan Africa, water scarcity during the dry season is a major problem, yet rainwater falling in the wet season is seldom collected except in large reservoirs serving those with piped water or in small water butts. In this case study, a rainwater harvesting system was designed with the objective of capturing and safely storing enough water locally, providing accessible water during the dry season while pupils continue to attend school and have water for drinking, sanitation and crop irrigation.

A number of key issues were identified when studying rainstorms and talking to local people in various Sub-Saharan African countries:

- Off-the-shelf guttering, where it exists, is not suitable for corrugated roofs with a large overhang common on African buildings.
- U-shaped gutters are quickly overwhelmed with the volume of water in a rainstorm.
- Expensive large-bore pipes are needed to rapidly remove the large volume of water falling on the roof during a storm from the storage tank.
- The storage tank must be large enough to hold sufficient water for the dry season; however, blow-moulded tanks currently available cannot be made large enough or be easily transported.
- The water tank needs to be enclosed to prevent insects from living and breeding in it and spreading disease.

A school in Zimbabwe piloted the devised system. It comprised wrap-around flexible gutters guiding rainwater from the roof into buffer tanks, where it was held, allowing the water to move under gravity via small-bore pipes to a large in-ground storage tank.

DOI: 10.1201/9781032638102-15

PROBLEM STATEMENT

In Sub-Saharan Africa, the lack of water during the dry seasons impacts many areas of life, yet rain that falls in the wet season is seldom collected. Western style guttering is unsuitable for use on many African buildings, so a new style of gutter for corrugated tin roofs was devised. The water collected was stored in large, enclosed tanks, preventing the ingress of dirt and insects and providing sufficient volume so that water was available throughout the dry season. Having water readily available during the dry season allowed the school to retain both pupils and teachers, preventing disruption of education.

The primary beneficiaries of the project were the pupils and teachers at Mwoyoweshumba Primary School in Mutasa district of Zimbabwe. The school has 400–500 pupils and teachers and has struggled with water access since it was established in 1973. The only natural water source is the River Nyamuchiri, which is approximately 6 km from the school. This water source is not suitable not only because of the distance from the school but also because there have been multiple fatal accidents at the river from students slipping on the rocks and falling in. To tackle this problem, a water pump was installed at the school, but it stopped working due to a continual fall in the level of the water table.

The lack of water access caused pupils to miss education as they needed to collect water. This in turn meant that Teachers did not stay long at the school, especially as they were also expected to collect water. Female pupils benefited the most from this project, as they are the ones expected to collect water and thus the ones who lose the most schooling.

ACTIONS TAKEN

A pilot project was installed at a school in Zimbabwe on a classroom roof. Local rainfall figures were used to calculate a predicted rainfall volume of over 850 m^3 on the classroom roof area in the rainy season. The heaviest rainfall was 12 mm in a day in a single storm. This equated to a volume of 1,680 L of rain collected off the roof.

Innovative wrap-around polyethylene gutters were attached to the edge of the roof. These collected the rain that fell onto the roof. The water then ran to each end of the gutter, where it was directed into buffer tanks through large bore pipes via collection vessels improvised from plastic bottles. These joined the gutters to the down pipes and quickly removed the water from the roof area (Figure C.S.5.1).

The buffer tanks were sized to accommodate a storm's worth of rainwater. The tanks were locally sourced and mounted on raised platforms, allowing the water to leave the tanks under gravity without the need for pumps. The outlets from the buffer tanks were connected to a smaller locally bought plastic water pipe, which led to the rainwater storage reservoir (Figure C.S.5.2).

The storage tank was placed in a trench in the ground, which was dug using local labour to fit the specification of the tank. The tank was sized to hold the rainwater collected from the classroom roof and from the top of the rainwater storage tank itself (Figure C.S.5.3).

FIGURE C.S.5.1   Wrap-around gutters and collection vessels on roof.

FIGURE C.S.5.2   Buffer tanks.

The tank and guttering were supplied as a flat pack, with all other parts of the system bought locally. The system was designed to be installed using local labour only to minimise costs and provide employment.

RESULTS

The pilot project was successful. Firstly, it was installed by local labour using the installation manual provided. There were slight deviations from the instructions, some of which were improvements to the design.

FIGURE C.S.5.3   Storage tank in trench.

Secondly, in the months between December and March, when most rain fell in the region, approximately 28,000 L of water were collected. This was a <90% collection efficiency based on the amount of rain that fell in the rainy season. The school was hit by Cyclone Idai in March 2019, but the rainwater harvesting system proved to be robust and sustained without any significant damage (Figure C.S.5.4).

Approximately 80% of the water collected was used in the dry season between April and October 2019. This allowed the school to remain fully open during that season (Figure C.S.5.5).

The most challenging areas were the importing of goods such as water tanks, pipes and fittings into Zimbabwe and the gathering of feedback. The importation problems were mitigated by using as much locally bought material as possible and reducing imports to the minimum. To obtain the feedback, a locally based organisation had to visit the school in person.

The lessons learned included the need to provide clearer instructions and more pictures in parts of the installation manual and to emphasise the need to treat the water collected if it was to be used as drinking water.

SUSTAINABLE DEVELOPMENT GOALS AND IMPACT

The project has impacted several SDGs, but particularly addressed the following:

SDG 4: Quality education
SDG 5: Gender equality
SDG 6: Clean water and sanitation
SDG 9: Industry, innovation, and infrastructure
SDG 10: Reduced inequalities
SDG 13: Climate action
SDG 17: Partnerships for the goals

FIGURE C.S.5.4   Storage tank full of water.

FIGURE C.S.5.5   Pupils using water from a rainwater storage tank.

The new system not only provided a steady supply of clean drinking water but also water for hand washing and toilet cleaning, thus having a direct impact on sanitation, health and wellbeing.

Not having to go to the river to collect water meant there was an increase in education quality as pupils could spend more time in the classroom and teachers were readily available to tutor students.

Having water available at the school has a great impact on girls, and it promotes gender equality. It is a societal expectation that girls fetch water whilst boys continue their education. The safety of pupils improved as they no longer had to collect water from the river, which was a dangerous activity with fatalities often reported. The rainwater harvesting system had an impact on local employment as it was installed using local labour.

System components were purchased locally where possible, supporting the local economy.

Water availability also had a positive impact on the psychological welfare of students and teachers, enabling them to focus on education without worrying about water access. Students no longer need to travel more than 5 km to the nearest water source. As such, they no longer miss out on classroom activities and suffer from tiredness from the long walks.

The system also impacted pupil and teacher numbers. After the rainwater harvesting system was installed, student enrollment rose by 35%. The headmaster believes this increase stems from the availability of water at the school. In addition, the school also managed to retain teachers. Previously, many stayed less than three years because of the water problems they faced.

NEXT STEPS

As this was a pilot project, it is expected that there would be modifications and improvements to subsequent installations. The feedback obtained from staff at Mwoyoweshumba Primary School and Zimbabwe Africa Trust, the charity that supports the school, will be used to make modifications to the design of the rainwater harvesting system to make it more robust and to improve the installation and functionality of the tanks. These include improving the method of holding up the tank in the trench, ways of putting the water pipes into and out of the storage tank, and covering the tanks to improve health and safety.

Bolsan and SOWTech are looking to become partners in a supported local manufacturing project to begin to build these systems in Zimbabwe. Zimbabwe Africa Trust has indicated that they have 51 more schools that could be interested in the system. There is also the possibility of extending the installation of these tanks to support small-scale farmers in Zimbabwe. A market analysis has been carried out on this consumer group, and it has shown an appetite to acquire such a water resource. Bolsan and SOWTech are continuing to pursue this new business enterprise.

ADVICE FOR OTHERS LOOKING TO DO SOMETHING SIMILAR

Do your research thoroughly as to the perceived problems as well as the actual ones, as these might not be obvious and can be a major stumbling block. For example, there was a perceived concern about carbon dioxide levels in the water, but carbon dioxide can be easily broken down by the body, and it is not a WHO-listed contaminant. Whereas there was an actual problem of protecting the children from falling onto the storage tank.

Find a good partner on the ground that can be your liaison between the project and yourself.

Listen to what your partner/client is saying. If they are not onboard 100%, they will not take ownership of the scheme, and it will be abandoned at the first sign of a problem.

Be prepared for things not to be done as expected. This may be due to miscommunication, unclear instructions or the installers deciding to do it another way. However, be open to changes, as they might not be negative and could simplify/improve things.

FURTHER INFORMATION

The inter-seasonal rainwater harvesting scheme was part of a larger project to design sustainable living equipment that could be manufactured locally in Sub-Saharan Africa using the minimum amount of imported materials. It was designed to be simple to install using unskilled labour.

Of paramount importance was safety. The water collected had to be as clean as possible and free of disease. The outlet from the buffer tanks was above the bottom of the tank, so any heavy material washed off the roof would be trapped at the bottom of the buffer tank and not pass into the main storage tank. The tanks were removed from the system so debris could be emptied out and the tanks cleaned. The storage tank had to be as enclosed as possible so that creatures could not get in, breed and lay eggs, causing the spread of disease. The water in the tank was not recommended for drinking unless it had been treated first.

On the SOWTech website (Rainwater harvesting (sowtech.com)), there is a short video of the school's headmaster and a teacher discussing the impact and benefits of the system, as well as a fuller report.

Part 3

Blueprint for the Future

9 Appropriate Technology in Rural Developing World Contexts

Designing for Simplicity, Durability, and Long-term Functionality

Harry Chaplin and John Gould

9.1 INTRODUCTION

Rainwater harvesting is an ideal solution for improving clean water access across a variety of contexts, cultures, and climates because it ticks so many boxes: simplicity, durability, and efficiency. It does what it says on the... tank? Rain – roof – gutter – storage – drinking glass – Sustainable Development Goal 6! However, even with rainwater harvesting, key considerations in design need to be adapted for the specific context in order to ensure that the system is simple, is durable and does the job it was installed to do in the first place. The main issue with so many water systems is dysfunctionality: no clean water, and in some cases, no water.

Water infrastructure in developing countries is plagued with functionality issues. Sustainable management systems are often lacking or non-existent and the wrong technology is installed in the wrong place at the wrong time, which exacerbates insufficient management structures. As a result, water systems often quickly fall into long-term or permanent disrepair. Despite significant initial capital investment, these systems are no longer able to provide water in sufficient quantities and/or at the quality standards required. Billions of dollars are spent on water infrastructure in poor countries each year by governments and NGOs, yet a huge portion of this is wasted as systems quickly stop functioning.

The longevity and efficacy of the management of a system come down to the practical ability and capacity of the management team to carry out the necessary monitoring and repairs. Simply put, if a highly complicated, highly energy-consuming desalination machine is installed in a rural area and there is not a specialist on hand to tinker with the machine over time to keep it running, the system is likely to fall into disrepair. If a low-tech, gravity-fed rainwater harvesting system is installed in the same community, there might still not be the basic plumbing or masonry skills

DOI: 10.1201/9781032638102-17

available, but it is likely that someone can fix the leaking tap. Assuming it is afford-able, a desalination plant would only be an appropriate technology selection where there is a highly skilled specialist, replacement parts are available, and there is suf-ficient power – and no other options are available.

Therefore, the choice and design of water supply technology should be specific to each situation. There is no one-size-fits-all solution. Applying the same design to multiple places at the same time on a large-scale leads to breakdowns in service. This is what makes scaling one solution successfully for long-term functionality so dif-ficult. It cannot be overstated how critical it is to select the right technology and the right infrastructure for the right context. Therefore, to build a well-performing and lasting water system, it is important to keep it simple and do the basics well.

- Step 1: Understand the context, including the target audience, community, users and the environment (social, political and geographical).
- Step 2: Working with the target users, select a solution that avoids unneces-sary complexity.

Given how large the world is and how large the demand is for clean water, why are there still hundreds of millions of people without clean water around the world? This book and this chapter can go some way in discussing and highlighting examples of organiza-tions and ways to scale via proper and appropriate technology selection and design.

9.1.1 What Is an Appropriate Water Technology?

Installing the correct type of water infrastructure and designing for the context is absolutely the first and most critical step towards the ongoing functionality of water infrastructure in rural areas. We call this appropriate technology, where the level of complexity for effective system operation and management is equally balanced with the capacity found locally to continue to operate it. We can also define this as being *in-step* with the community. The best way to create a solution to a water issue that is in-step with the community's capacities and needs is to always design it in collabora-tion with the community itself or the community involved.

Local capacity can be the skills needed to change a broken pipe; it can be the availability of the replacement broken pipe; or it can also be the financial ability to purchase and fit the broken segment, inclusive of materials and salaries.

It is the combination of appropriate technology design and installation with ade-quate local capacity that leads to long-term functionality. In this chapter, real-life examples and situations of successful and failed water infrastructure are presented, and rainwater harvesting is discussed as an appropriate technology for rural areas.

9.2 WATER SYSTEMS IN RURAL AREAS: PERSPECTIVES FROM RURAL MADAGASCAR

To provide some context, roads in parts of Madagascar are peppered with large rocks and large potholes and can turn into flowing rivers themselves in torrential rain-fall. It could take an hour to travel a kilometer. In eastern Madagascar, roads are punctuated by rivers spanning 20–400 m across. The metal barge used to transport a car one at a time is often leaking and needs bailing out before and during the crossing. Sometimes the motor is broken, and so the entire six-tonne cargo may need to be pulled across by passengers and barge personnel. If the rope snaps, the community might help out by emerging in force, 40 people strong, to paddle you and your massive load across the swollen river with handmade wooden paddles. The quicker and safer mode of travel is via motorbike. It can be satisfying to drive past the queue of 100 cars and trucks waiting to do a 400-m river one-by-one and go straight across with your moto in the *pirogue (dugout canoe)* with you – literally saving days of travel time. However, it is difficult to carry bags of cement and 6-m PVC pipes and fittings off-road on a motorbike.

Remote rural communities (e.g. Figure 9.1) can be difficult to get to, taking days if not weeks to travel. Choosing the wrong technology at the design phase here will almost definitely result in dysfunctionality. It is crucial that the designer understand the local context from all angles, and if this is not possible, then to have at least the time and the financial capacity following installation to monitor, evaluate and adapt the system installed. In these remote corners of the world, low-tech is most often better than high-tech. However, the rate of development in water technology is rapid, and one must also acknowledge when a new technology, e.g., a filter, is no longer a gimmick but is a feasible, long-term solution to improving water.

FIGURE 9.1 Ecole Publique Primaire Mananara II (primary school) in the mountains of Madagascar takes 2 hours by foot or by motorbike on some very difficult and wet tracks – a 4×4 car cannot access this community.

9.2.1 "PERCEIVED COMPLEXITY"

Imagine the scene. You are the WASH specialist on an NGO project, and you have been working in the sector for 20 years across multiple countries. You have arrived at a mountain village after two days' drive on off-road conditions. You are actually only 113 km from the nearest large town, but it has taken you 25 hours to get to the village. Your pickup 4×4 has broken down four times, and you had to wait at one river crossing for another 3 hours for the water level to subside so that the barge could be floated again. In the back of your vehicle, you are transporting 20 new simple sand filters that your NGO has made back in town, and you are coming to carry out a sensitization of the community. Perhaps some CLTS shock tactics will be used. (Community-led total sanitation is often used to attempt to "shock" people into realizing certain habits are causing illness within a community.) You want the community to really understand how these filters will change and improve their water. You have planned multiple other community visits on the same field mission, so you get straight to work once you arrive in the village.

At the end of the day, your CLTS-triggering sensitizations are finished. Tick. Your capacity-building activities have been carried out. Tick. Your report is going to look amazing, filled with photos of your team in their donor-branded T-shirts lined up with beneficiaries from the community. You feel great, and the community is smiling because you've given them 20 new filters. The donors are going to be happy and fund you to go to the next region next year to do the same one-day visits. Win-win-win.

Two weeks later, you arrange a call with the community committee that the Chef Fokontany (village chief) elected: It actually turns out, unbeknownst to you, that this group of people is also the committee for the local agricultural project from another NGO, the revolving fund scheme from a third project, and the leaders of the decision-making team for the fishing project from a fourth NGO, despite being in the mountains on a three-day hike from the ocean. Out of the 15 people in the committee, 13 are men, and all but four of the entire committee live permanently in the large town that you travelled from. They also only came to the village that day for the training and to receive the filters. The Chief, absent on the day of the visit, announces that all the filters have clogged up, that no one is using them anymore and that they have returned to collecting water from their original, unprotected sources. Training was given to backwash the filters, and you try to coach the committee over the phone. Most of the telephone numbers aren't working, likely a lack of battery power or signal, and when you finally do make it through to one of the committee members, they are not in the village but in the same town as you.

You reschedule the call, and after five more times, you finally get through to someone on the committee who is in the village, has a battery on their phone and has a signal to receive the call. You explain the same steps as on the training day, and the committee member understands what they need to do to get the filters working again. Change the sand or flush the sand in a bucket, nothing particularly complicated, you think. Two weeks later, and the filters still aren't being used. The committee member you spoke with had a death in the family and left the village to be with their extended family. Upon return, they were busy with the new rice harvest. The filters are still not being used.

The project funding finishes, and final reports are submitted. So there is no means to return to the community to address the issues, so the filters sit unused just one month after deployment.

Now imagine the other side. You are from rural southeast Madagascar, a one-day walk (or two-day drive) from the nearest large town. You live in a small village in a bowl of mountains and live mostly off your own farming for survival. You grow rice and a few vegetables. You have some lychee and some mango trees, which earn you some money at harvest time once a year. However, since all your neighbours also have lychee and mango trees, the price at the market is low, and you do not have the transport to reach the large town where you can sell for a higher price.

You currently collect water from a nearby well in your village, but the pump head has long since broken, so the top is open to the air, and everyone collects water with a bucket and rope. The water at the bottom is dirty, but you have noticed that drawing off from the top usually minimizes the dirt you can see in the water. The colour is clear, the taste is okay but earthy, and only in the really dry periods does the water start to smell. You and your family, especially your three young children, get stomach pains and diarrhoea periodically. You have soap at home and a dug toilet. You know that it is likely that the water is the cause of your stomach pains and diarrhoea but you continue using it regardless because there is no better alternative.

One day, you hear a car arrive in the village and find an NGO project team getting out of the car. One of the team members is a foreigner not from Madagascar, and the other two are Malagasy (a term used for people from Madagascar and also the name of the local language) but clearly from another region in Madagascar because their accents are different and they use some different vocabulary that you find it difficult to understand, and the last member of the team is a Malagasy person from the local town. They explain that they are there to distribute water filters and carry out training about clean water and WASH practices. The village chief is not in the village, but the committee that he selected says that they had organized a community meeting for that day with him. Some committee members are present, and about an hour later, roughly 30 people have turned up. It is rice harvest season, so many of the women are not present; the group is mostly men and teenage boys.

The NGO team started with an exercise about drinking water and how the colour is not necessarily linked with how clean the water is; they even dipped a single hair in a pile of poo and then afterwards in a clear glass of water poured from a new bottle they brought from town. The colour did not change, but still, no one would drink it afterwards.

They then talked about the 20 water filters they brought in the car and explained how they worked and how to wash them if they became slow or stopped working. Many people from the community who were watching seemed interested, and after the NGO left, the filters were distributed and used a lot for two weeks. After a month, the water stopped flowing out of the filter, and those with filters came together to figure out the next steps.

The main barrier was that the filter looked complicated, and there was apprehension about removing the sand and gravel layers because of fear of the NGO's strict instructions. Few of the committee members present were technically minded and understood the demonstrations on backwashing and changing the sand and gravel.

After all this hassle, you decide the water coming directly from the well is now better than that coming out of the water filter and abandon the water filter.

Trying to experience the situation from the community's perspective is equally as important as proper system design and technology selection. One day is clearly not enough to fully integrate the new filters into the daily routine of the families in the village, and there are no lines of effective communication available for ongoing support beyond the project's end.

It is likely that a sand filter in this context would have been a beneficial solution to the communities' water quality issues, but since the design and the building of the technology were not made in collaboration with the community, there is no long-term understanding and thus no functionality. If the project cycle had not ended and the model had used an approach where there was funding for a second, third, etc. visit to the community, the integration of the filters into the village could have perhaps succeeded.

Users of technology in a rural, developing context might perceive the technology as complicated, no matter how simple the technology might seem to the original designer or project team. This perceived complexity can lead to users becoming demotivated to manage water technology properly and even abandoning the technology completely the moment it becomes difficult or perceptively difficult to manage or maintain. Effective communication, contact time and technology selection appropriate to the target context can overcome this perceived complexity. Even better is to establish communication links beyond the life of a project to the original organization or government officials that had installed or donated the technology in the first instance.

In rainwater harvesting, there are several elements of the system that can be added to and taken away to make the system more efficient in the collection, storage and distribution of clean water. The next section examines some examples of these specific design features with respect to various contexts and discusses how appropriate they are or not for long-term functionality.

9.3   CONSIDERATIONS FOR RAINWATER HARVESTING SYSTEMS IN RURAL AREAS

9.3.1   System Design

Roof → bucket.

The ultimate simple rainwater harvesting setup. However, it is not effective in quantity or in quality. The proportion of rain falling on the roof that goes into the storage tank is called the runoff coefficient, denoted by the letter c in the following equation:

$$V = PAc$$

where V is the volume of water in litres collected from a rainfall event, P is measured in mm falling on a surface area and A is measured in m^2. Following this equation, we can simply remember that 1 mm of rainfall falling on 1 m^2 of roof collection surface gives approximately 1 L of water. Obviously, if there is no gutter, we're only collecting on the strip of collection surface running into the bucket. Likewise, we will clearly lose a significant amount of rainfall if the bucket is at ground level, 3 m below the roof eave!

Material selection for the roof affects the volume collected. A typical tin roof has a $c = 0.8$–0.9, i.e., 80%–90% of the rain falling on the roof is collected into the tank. A grass roof is nearer to 20%–30%.

The gutter also plays an important role in both quantity and quality. A gutter that spans the entire length of the eave of the roof and that has a sufficient gradient – usually a drop in height of 1 cm for every 1 m in length is sufficient – will ensure maximum collection, minimum splash losses and quick conveyance into the tank. A flat gutter might well avoid all splash losses by being very close to the roof eave for the entire length, but it also leads to water stagnating in the gutter. Water sitting in an open gutter exposed to the sun is at risk of contamination by birds, dust and dirt blown into the gutters or onto the roof during dry periods and generally provides a place for bacteria to thrive. Sitting water also provides an ideal breeding environment for mosquitos, which is a risk in malaria areas.

Even if the gutter is positioned correctly at a good gradient for quick conveyance of water, there is always the risk of bird droppings, dust and sand being blown onto the roof and other debris being deposited onto the roof during a dry period. It is imperative to avoid these contaminants being transferred into the storage tanks, thus contaminating the water stored from previous rains.

The options would be a series of obstacles to divert contamination away from the storage tank. Firstly, we install a leaf mesh (Figure 9.2) that is effectively a 1–2 mm spaced mesh that is placed after the gutter to catch all large debris from entering the final downpipe. The risk here is blocking the top of the down pipe with accumulated leaf or other debris from the gutters or roof and subsequently losing water, dramatically decreasing c. There are several different designs for a leaf blocker/mesh/eater that minimizes this risk. The most frequently used method is simply inserting the mesh at an angle, and thus the debris is washed off by the next water and the mesh self-regulates. However, there is always a slight decrease in c when using leaf meshes.

Following the leaf mesh, a first flush mechanism (Figures 9.3 and 9.4) can divert the initial rainfall after a dry period away from entering the tank. This initial rainfall event washes the roof and gutters of any contamination, protecting the high quality of the rainwater.

Some rainwater systems might include a filter for pre-storage. The risk here is that stored water can be more easily contaminated unless the system is bespoke and built at a very high level of craftsmanship or using perfectly fitting (usually more expensive) factory-made parts and closely monitored over time.

A fine mesh might be installed before the downpipe delivers water into the storage tank. Research shows that a dark and entirely closed water storage tank is conducive to microbe decline since bacteria are starved of new contaminants and sunlight to feed off. For a tank with a closed roof and the downpipe entering through a small 100-mm opening or a manhole-sized opening, this mesh serves two purposes: to prevent any large debris that might have bypassed the first flush mechanism and to seal the storage tank.

The downpipe entering the tank can also be modified to ensure optimum water conveyance. Instead of depositing water into the top of the tank at ~2 m height and

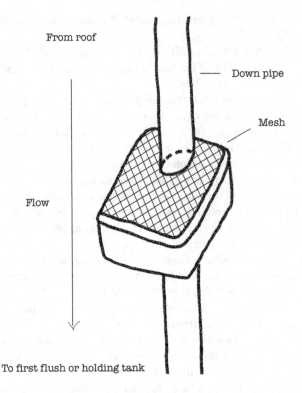

From roof

Down pipe

Mesh

Flow

To first flush or holding tank

FIGURE 9.2   A typical "leaf eater," otherwise known as a "gutter snipe," traps large debris and prevents it from entering the downpipe to improve runoff.

dropping to the water surface, a downpipe can be inserted that extends to the bottom of the tank. This downpipe delivers water directly to the bottom of the storage tank, preventing major disturbance of any sediment or other contaminants in the tank and allowing for continued clean water draw-off by users. A protective footing can even be installed at the bottom of this downpipe to protect it from disturbance to an even greater extent.

Inside the tank, we should avoid drawing water from below 100 mm from the bottom of the tank to avoid any sediment draw-off. This can be a fixed pipe that rises 100 mm or a flexible floating pipe to draw from the top of the water level. The latter is used in Isla Urbana's rainwater installations in Mexico, where a flexible pipe inside the tank is connected to the draw-off pipe at the bottom of the tank. A floater is attached to the end of the flexible pipe and ensures that water drawn from the tank is always taken from the highest water level in the tank – at the farthest point from sediment settled at the bottom of the tank. Isla Urbana takes it a step further and connects a small, spaced stainless-steel meshing at the end of the draw-off pipe to avoid any floating debris from entering the draw-off.

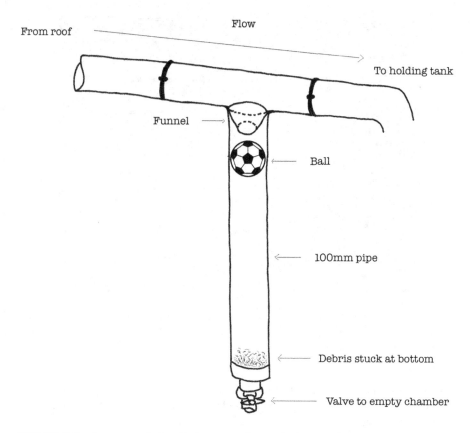

FIGURE 9.3   An automatic first flush system diverts the first rainfall into the vertical pipe whilst the next rainwater flows into the tank.

A draw-off tap can be connected directly to the tank, piped under pressure (using an electric water pump), or via gravity to a water point further from the tank.

The importance of the role of effective end-user management in the story of water quality should not be forgotten or understated. A rainwater system can provide perfect-quality water at all international standards, but a contaminated water collection vessel can quickly and effectively make all those efforts redundant. Therefore, it is also essential that water collection vessels are cleaned properly before filling them with clean water.

So, we started with Roof → Bucket and now we have options layered on: Roof → well-positioned gutter → wide-spaced and angled leaf mesh → first flush diversion with floater ball → inlet mesh to the closed roof tank → flexible pipe with fine mesh to draw off water at the top of the tank → gravity-fed water point external to the tank wall. Don't forget the specific material selection available at each stage and the advantages and disadvantages of each – again context specific.

In coastal areas, the roof and other metal fittings should be protected from rusting. Gutters should be selected for ultimate longevity. Galvanized metal gutters may

FIGURE 9.4   Ecole Publique Primaire EPP Ambinanikely (primary school) in Fort-Dauphin, southeast Madagascar, shows the placement of a vertical first flush between the gutter and the water tank.

rust, but then the direct sunlight may deform the plastic PVC gutters. A leaf mesh installed in a really remote community might prove too much risk for effective and simple management. What if it becomes blocked with each rain and requires intensive cleaning by the agent, which means it loses a lot of water each time? The first flush floater ball mechanism with a vertical pipe could be a brilliant add-on to protect rainwater quality before the tank, but it requires emptying before the next rain for it to be functional, and that relies on a motivated person doing this. The inlet mesh to the tank is equally a simple but effective measure for ensuring the crucial last barrier to prevent debris from entering the tank, but it also needs rinsing and washing periodically to avoid contaminating all runoff. A flexible draw-off pipe floating to the top of the tank provides an additional moving part to break but given the correct training and capacity to monitor this, it could be a significant step in ensuring the cleanest water at all times.

Where possible, a gravity-fed water point is always preferable to one that needs an electric pump. As soon as electricity is required, solar panels or other consumables (fuel for a generator) add to the risks of breakdown and theft, which would then render the entire system dysfunctional. A tap is used tens or hundreds of times a day and, crucially, moves. This relatively inexpensive piece of equipment, whilst easy to install by most people, can lead to tens of thousands of dollars' worth of infrastructure not being used if it is broken. A loose tap in the side of the tank can lead to losing the entirety of the water stored, whereas simply installing it at a water point 2 m away that is piped by gravity from the tank avoids this risk. The tap will likely break but can be replaced. A stop valve between the water point and the tank will prevent the water stored in the tank from being completely lost.

FIGURE 9.5   A 5,000-L calabash tank constructed in Guinea-Bissau.

Rainwater harvesting systems can be diverse in their make-up, and depending on the local context, they can be complex and appropriate or simple and appropriate. In Madagascar, given the remote and rural nature of the water systems, the following setup may be the most suitable: roof → gutter → ball floater vertical downpipe first flush with a 40 mm stop valve at the bottom for resetting → coffee filter cloth over the manhole entry to the Calabash tank (see Figure 9.5) → and a draw-off pipe at 10 cm above the base of the tank feeding water points by gravity and a second, larger diameter pipe at 0 cm above the base for cleaning and flushing of the tank every 3–6 months. Water quality tests over the years have proven it is possible to obtain international WHO standards given proper installation and management of rainwater systems – even without filtration or chlorination.

9.3.2 Appropriate Water Storage Design and Installation

Water storage is the most expensive component of a rainwater harvesting setup. There are a multitude of options to choose from. Factory-made plastic tanks are portable, uniform in production and quality, and, in many places, the most economical solution. For communities connected to production hubs, these plastic tanks are likely the most appropriate storage solution. However, in a remote rural locale, transport of plastic tanks can be very expensive, and upon finding a leak in the system, capacity in materials, finance to purchase materials and/or skills to do the fix might not be available.

The calabash tank is a type of ferrocement-built water tank modelled on the calabash fruit in Guinea-Bissau in 2010 by Paul Akkerman and his team (Akkerman, 2023). The simplicity and efficiency of the tank have led to its adoption across sub-Saharan Africa and Latin America. One of the biggest benefits of building calabash storage tanks is their replicability and scalability with local knowledge and masonry skills. It is appropriate in many situations because of its accessibility in price and because it can be built in-situ rather than transported. In some ways, it is the perfect example of how technology is designed and built in step with the capacity of a community. In this case, the calabash design can fit many different contexts because of its simplicity and widespread applicability.

Another example of appropriate water storage is the widely reported Thai Jar example from Thailand (Figure 9.6). In the 1980s, the Thai government subsidized the sale of portable 1,000 and 2,000-L water jars for rainwater collection to incentivize private production and boost the number of households accessing rainwater harvesting at home. Between 1981 and 1992, more than 8 million jars were constructed, three-quarters of which were thanks to government subsidies. By the year 2000, over 50% of the rural population of Thailand (21 million people) were using rainwater harvesting as their primary drinking water source (Saladin, 2016). This led to Thailand having the smallest difference in access to improved water between the richest and poorest quintiles among middle- and low-income countries worldwide. This is an example of an effective, appropriate water storage solution for homes that is completely understandable, relatable, and scalable by regular, non-specialist people. Here is a user-friendly manual that Tatirano produced replicating the Thai Jar in Madagascar (see further reading).

9.4 DISCUSSION

Ultimately, appropriate technology in rural developing contexts for long-term functionality depends on alternative water sources available to communities and the acceptability by communities of both the water quality and the water technology producing this water.

Water access can be represented as a ladder with a series of rungs. Communities at the bottom of the ladder with only surface water open to contamination are at the bottom rung. An urban settlement with drinkable piped supply into the home is at the top rung. The appropriate design and technology selection should account for which rung on the ladder the target community is on. Smaller steps up the rungs of the ladder, improving the water provision and aiming for the WHO standards eventually is a far more realistic and appropriate strategy.

When seeking urgent solutions to water supply provision, incremental improvement may be the only practical approach. For example, providing water supplies

FIGURE 9.6   Thai jars from Thailand are very low-cost and portable. Tatirano has successfully replicated these jars in Madagascar, forming the Madagascar Jar.

that meet WHO microbial standards in a remote community in the mountains of Madagascar may be inappropriate, especially if there is little evidence that the community's health is being compromised.

9.5 CONCLUSION

Rainwater collection is a fantastic example of an appropriate technology for application in rural, developing-world contexts. It encompasses the philosophy of being easily understandable, easy and affordable to construct and easy to maintain over time. When properly maintained, rainwater collection systems can provide a perfectly clean drinking water source at WHO standards even without filtration or treatment.

FURTHER READINGS

Gould, J. and Nissen-Petersen, E. *Rainwater Catchment Systems for Domestic Supply: Design, Construction and Implementation*; 1999. Intermediate Technology Publications: London, 335 p.

International Rainwater Catchment Systems Experiences: Towards Water Security, May 2020, IWA Publishing. https://doi.org/10.2166/9781789060584, https://iwaponline.com/ebooks/book/790/International-Rainwater-Catchment-Systems

Madagascar Jar Build Guide by Tatirano Social Enterprise. https://drive.google.com/file/d/1y
    MZJLkfCZPEgxXTBl1kFuGHeXQU2q-iQ/view?usp=sharing.
*Rainwater Harvesting in the Homestead, a series of ideas by Peter Morgan, the 2013 winner
    of the Stockholm Prize for Water. Peter Morgan is also a leader in the sanitation space,
    having designed and developed some of the first improved pit latrines.* The above hand-
    book can be accessed at https://www.rural-water-supply.net/en/resources/details/666
Rural Water Supply Network – Rainwater. https://www.ruralwatersupply.net/en/search?search=
    rainwater
Thomas, T., and Martinson, D. *Roofwater Harvesting: A handbook for Practitioners.* Technical
    Paper 49, IRC International Water and Sanitation Centre: Delft, Netherlands; 2007.
Water for Aridlands, A Collection of Articles and Videos Created by the Late Erik
    Nissen-Petersen. Some found here: https://www.samsamwater.com/library.php?serie=
    Water%20for%20Arid%20Lands

REFERENCES

Akkerman, P. 2023. *Rainwater Harvesting in Africa - Calabash Cistern Manual 2023.* Netherlands.
    Accessed 08/02/2023. https://degevuldewaterkruik.nl/assets/uploads/pdf/2023/calabash-
    manual-2023-en.pdf
Saladin, M. 2016. Rainwater Harvesting in Thailand: Learning from the World Champions,
    Rural Water Supply Network. Accessed 08/02/2023.https://www.rural-water-supply.net/
    fr/ressources/details/759

10 Rainwater Harvesting in the 21st Century
Impactful Water Partnerships

Karl Zimmermann

10.1 INTRODUCTION

10.1.1 ON TRUST

Everyone has people and things that they trust more than others. Perhaps someone can trust an office tower in Canada not to collapse; trust is placed in the unknown engineers who built the building. Conversely, some people will not drink chlorinated water; they may not trust health practitioners that this toxic chemical can be good for their health in controlled doses. So, what is it that lets someone trust one group of people but not another?

Trust is *pro*-spective: through a risk-based assessment of someone's abilities (which are unknown to the decision-maker), one may decide to *trust* an unknown engineer's design abilities and climb eight floors to an office. Conversely, 'confidence' is *retro*-spective: based on an existing track record of a century of construction projects, do we have *confidence* in the use of reinforced concrete to hold up the eighth floor? Generally, we would prefer to have confidence based on a shining track record. But as described by Enang et al. (2008), in cases where there is insufficient past evidence for us to predict future success, we must rely on a 'leap of faith' into the unknown, on trust. This is often the case with rural water systems, especially for the first generation in a community who has access to improved water sources.

Most people have not lived through a building collapse but have likely heard of someone who got sick from contaminated water, especially if living in a low-income area. And so, the track record of drinking water systems has blemishes, straining people's abilities to rely on confidence. Few people know the details of a breakpoint chlorine calculation or the mechanism of chlorine's toxicity to enteric pathogens. In fact, some people would consider a water filter's 99% inactivation rate of pathogens a marvel, while a water operator may consider this 2-log performance insufficient. Because of this uncertainty, there could be a lack of both confidence and trust in chlorinated drinking water. Consider how many of the things that people often *don't trust* are the result of their lack of understanding. Whereas how many of the things that people do have *confidence* in are because of a successful track record? It is important to consider history and existing knowledge and experiences when proposing a safe water solution.

DOI: 10.1201/9781032638102-18

10.1.2   A History of Empowering End-users

The essence of the Water Partnership approach is trusting water users with decision-making roles. Water partnerships go beyond 'informing' and 'consulting' (forms
of tokenism, defined by Arnstein (1969)), towards community-led decision-making
power and control. While exploring the tools for community-led decision-making, it is
useful to understand the history of end-user empowerment. This section was informed
by Cornwall's (2006) excellent review titled *'Historical Perspectives on Participation
in Development,"* which details the emergence of participatory engagement.

Today, it is easy to criticise a 'top-down' approach to development work, yet it has
distant origins. Since the time of thinkers like Plato and Newton, there have been
beliefs that 'professionals know best'. Public views were discarded when a scientist
offered his 'educated' opinion; this is top-down thinking.

In the early twentieth century, Britain was considering self-governance solutions
for its colonies. A 1922 book by Lord Lugard introduced 'The Dual Mandate', giving certain powers to traditional leadership, which inspired forthcoming decolonial-
isation efforts. Over the next decades, critics argued of inequalities resulting from
domination by local oligarchs and little popular participation.

Following the Second World War, British colonialism in Africa generally sought to
transfer power to colonial-educated Africans. At the same time, the United States and
its allies observed the rapid post-war rebound of European economies. This inspired
the thought that poor countries were poor because they didn't have the resources and
technology to be rich; the solution was the provision of wealth and technologies!
State-led initiatives provided the means for poor societies to rapidly develop. This
was captured in the 1956 U.N. Economic and Social Council, which encouraged *"the
participation of the people themselves in efforts to improve their level of living with
as much reliance as possible on their own initiative, and the provision of technical
and other services in ways which encourage initiative, self-help and mutual help."*
However, here too, criticisms emerged that mandating participation was simply a
way to keep local people busy and suppress urges to revolt or dissent.

The later 1960s and 1970s shifted focus towards market-driven economic development while emphasising voluntary participation in development initiatives and
equitable sharing of their benefits. The US Foreign Assistance Act of 1966 encouraged *"maximum participation in the task of economic development on the part of the
people of the developing countries"* and *"[...] in this area of development activity,
both the initiative and the human resources should come primarily from aid-receiv-
ing countries."*

Next, the 1977 World Water Conference in Argentina officially adopted community participation for sustainable water management, coinciding with the 1980s
International Drinking Water Supply and Sanitation Decade. Later gatherings in the
early 1990s re-affirmed a commitment to participatory engagement and the transfer of ownership and responsibility for water management to the end-user level, or
as close to there as possible. This was in part interpreted by emphasising cash or
in-kind donations in hopes of leaving 'ownership' to be dealt with by the receiving
population. A reflection by Paul (1987) described an emphasis on cost-sharing and
co-production of services to constitute community participation, while a UNICEF

report further advocated that community participation "*is essential for the planning, implementing, and success of the approaches devised, as well as for keeping the cost of the programmes down by means of community contributions.*"

Many initiatives in the second half of the 20th century adopted an approach of importing foreign-designed (and in many cases, foreign-built) solutions for water, representing the height of the 'Import Model' for development (Figure 10.1a). The Import Model worked for the top-down view that poverty might be solved by importing resources and advanced technologies from rich areas to poor ones. Examples might include handing out boxes of water filters, showing up in a community to drill boreholes, or installing ceramic filters for community water systems. However, the sector learned that without the solutions being adapted to the local context, having supply chains for spare parts, and especially in cases lacking an understanding of water contaminants, their impact on health, and the need to always use the filter, *imported* water solutions often failed.

As the development sector learned the limitations of importing technologies, they also learned that local actors were critical to long-term success, and so a new paradigm emerged: what we may now call the 'Strings Attached Model' (Figure 1b). Through the 1990s, the focus shifted towards supporting community-level organisations (in many cases, NGOs) that possessed a combination of existing roots in a village, more cost-effective outcomes (due to less top-heavy organisational structures), and a more personable approach to working with end-users, leading to greater access to disadvantaged people. Instead of paying directly to purchase and deliver water filters, donors began to instead provide their resources to local organisations to carry

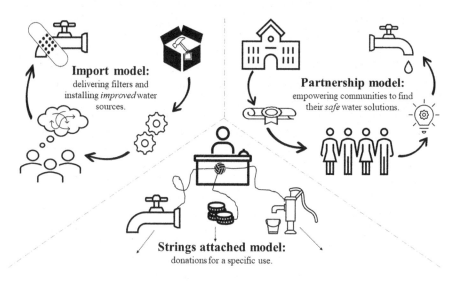

Import model:
delivering filters and installing *improved* water sources.

Partnership model:
empowering communities to find their *safe* water solutions.

Strings attached model:
donations for a specific use.

FIGURE 10.1 Previous development efforts focused on either (a) importing improved water infrastructure or perhaps (b) providing local organisations with resources but having their use restricted to certain objectives. More recently, the development sector is moving towards a water partnership model (c) where decision-making power remains close to the water users who seek and develop their local water solutions.

out largely the same programmes. However, the objectives, methods and oftentimes the water solutions are determined by a foreign organisation, with the local actors left only to carry out the programme without meaningful decision-making power. Resources are provided, but with 'strings attached'.

Towards the turn of the 21st century, the development sector began adopting rights-based approaches to justify its interventions, focused on providing the necessities of life where the state could not (or was not). This informed the UN's Millennium Development Goals in 2000, including Target 7c to halve the number of people without access to safe water and basic sanitation. By 2015, the Joint Monitoring Programme reported success towards Target 7c, with 2.6 billion people having gained access to improved water and 2.1 billion to improved sanitation since 1990. However, critics argued that the definition of 'improved' drinking water focused only on the infrastructure and not its management. It was observed that infrastructure by itself does not assure safe water Onda et al. (2012) and so the UN's Sustainable Development Goals (from 2015 to 2030) emphasise *safely managed* water and sanitation'.

However, more recently, this model of pouring financial support behind NGOs has received criticism, whereby organisations suffer a gradual erosion of their vision and morals in attempts to appease the interests of foreign supporters. Furthermore, the requirement of foreign donors for Western-style quarterly reports and financial accounting can create tensions with local business practices and divert resources from the tasks at hand (although accountability measures do indeed have a place towards ensuring sustainability).

And so, in the 2020s, the sustainable development sector has arrived at the need for a new model for participatory engagement. Here, we propose that the next paradigm in participatory engagement should involve ceding not only ownership but also the decision-making power of the water project to the beneficiaries. Importantly, this does not mean that there is no role for foreign organisations. Instead, the development sector today is learning the avenues for the other stakeholders to support the end users in their quest for safe water solutions. In other words, how to create a collaborative *Water Partnership*.

This chapter will explain that empowering end-user decision-making is best achieved through building partnerships based on trust. The discussion will continue by numerating recommendations to build water partnerships and the roles for each stakeholder. Water partnerships empower end users into decision-making roles, building collaborative teams where each partner contributes their expertise while also trusting the judgments of others. This strategy is appropriate not only for water projects but for many partnerships requiring relationships built on trust, especially for culturally significant topics like water, health, and livelihoods. The stories and lessons recounted in this chapter follow from discussions with water leaders in Madagascar while visiting Tatirano Social Enterprise, an NGO specialising in rainwater harvesting systems in rural southeast Madagascar.

10.1.3 SETTING THE SCENE: MADAGASCAR

As always, we woke at sunrise to start the day with a cup of sugar-laden coffee. It was too sweet for my Canadian palate but paired well with the deep-fried dough-balls, '*mofo*'. I was visiting with Tatirano Social Enterprise, and our team of six was returning from a two-week journey along the southeast coast of Madagascar, checking in with rainwater harvesting systems and their kiosk operators, and scoping future locations. For the next three days, our objective was to get back to the town of Fort Dauphin, just 160 km south. It was March and it had been raining. Our journey included ten river crossings using ferries, called 'bac' in Malagasy, seven of which are self-powered by pulling on a rope. Around 2:00 pm, we arrived at the Iabokoho River, which had swollen by six feet with recent rains. As we arrived, a Land Cruiser was disembarking the 20-foot ferry raft. It landed in four feet of water, sputtering the engine, but managed to drive away. We were told to wait; a crossing wasn't possible until the water receded.

And so, we waited. We shared a bushel of bananas; the water rose 15 cm. We read a book; the water rose 30 cm. We had dinner in the dining room of a local house; the water rose 10 cm but was slowing. The sun set with us still on the north shore and the hotel-like rooms impossibly close on the south. We did the only reasonable thing: we bought a bottle of rum and played poker in a mixture of Malagasy-French-English.

By sunrise, the water had receded enough, and the ferry operator piloted us on the four-minute journey across the fateful Iabokoho River. Once on the south bank, we sped away at 18 km/h down the bumpy road.

This is the reality of working in developing contexts. Often, the infrastructure you might expect is in disrepair or missing. It may take weeks to move a cement truck or construction crane around these regions, and a supply chain for replacement stainless steel bolts may takes weeks. If the project needs electricity, a generator should be arranged. If a weekly delivery of disinfecting chlorine is needed, it might be four days late. This means sustainable development requires finding solutions that work in the local contexts: localised solutions.

10.1.4 INTRODUCING THE WATER PARTNERSHIPS MODEL

When beginning to work with a new village, people don't know much about each other's backgrounds, and without a track record, there is a lack of confidence. Perhaps, someone is proposing a new water technology that is poorly understood, and so there is also a lack of trust in this approach. Here, we propose that creating partnerships built on trust *between the water team* carrying out the project minimises the impetus on water users to trust the technology alone.

Distrust can result when a foreign actor intends to *import* a water solution that has likely been proven in a research laboratory or even in the field in another location. Instead, here we advocate the potential of the *partnership model,* whereby we adopt the mindset that the end-users own their water challenges and their water solutions. In a Water Partnerships approach, the role of partners is to support end-users on their quest for a safe water solution. This approach also adopts a rights-based view whereby communities that may legally own their water resources and individuals who own their right to a happy and healthy life are empowered to increase their access to health and happiness. It is hence the role of the other water stakeholders to find the appropriate avenue to support end-users on their journey to a healthy and happy life through water solutions.

Thankfully, innovation comes naturally to a group of people who will ultimately benefit from its creation. In an essay on *'Democratising Innovation'*, von Hippel

(2009) explained that up to 80% of inventions for scientific instruments, licenced chemical production processes, and innovations in oil refining were in fact developed by people who were also the product's users. We extend this finding such that a successful water innovation is very likely to emerge from the people who have lived in a village with its seasonal water resources, its changing water needs, and an understanding of the past water successes and failures. In fact, one of the most powerful themes to emerge from this study reiterates the potential for end-user innovation: *'People understand their water challenges, and they understand their water solutions too.'*

For decades, the development sector has been moving towards participatory engagement and bottom-up approaches. I believe that we're now at a point where we understand the value of involving end-user voices. But the question remains, *"how might we best support the end-users to innovate their safe water solution?"* From the partnership model's perspective, what are the tools to build water partnerships? We will address these recommendations shortly.

10.2   METHODS

This study sought to understand the recommendations of diverse water sector actors towards creating sustainable and long-term solutions to safe drinking water by employing a partnership approach.

Semi-structured interviews were conducted with dozens of stakeholders, including the leadership and field staff of Tatirano Social Enterprise, government ministers, water sector actors from other NGOs, community health practitioners, teachers, and women standing in line at 6 am to collect water. Participants were purposefully selected from the network of water professionals connected to Tatirano Social Enterprise, a rainwater harvesting NGO in south-eastern Madagascar. Interviews were conducted in English, French, and Malagasy, where the latter were translated on the spot to English. All interviews were audio-recorded and later transcribed into English. Mixed inductive and deductive coding using NVivo revealed the network of partners and their roles, as described by themselves and others in successful past water partnerships. Interviews were conducted with home-university research ethics approvals.

10.2.1   It Takes a Village: Definitions of Water Partners

It is also worth explaining who are the 'water team stakeholders': everyone involved in supporting the community on their journey to safe water. This includes the water users themselves, village leadership, health authorities, teachers, local or national government, the private sector, NGOs, local and foreign technical experts, donors, and others.

It may also be of use to distinguish between a village and a community. The former represents a geographical grouping of families; a village can be circled on a map. Whereas the latter invokes social elements, whereby people may belong to many 'communities' related to their neighbourhood, social circles, livelihoods, past-time activities, etc. Villages and communities may indeed overlap, but not necessarily so. Some people may identify with their religious community or the community of artists within their village. In addition, the following discussion will use 'foreigners' to describe people who are distant from the local communities. Depending on their relationships, people of the same nationality may be considered foreigners to a particular community if they are not perceived to share similar values or have similar opinions and experiences.

10.3 FINDINGS

10.3.1 WATER PARTNERS AND THEIR ROLES

Let us consider who the important water partners are. Discussions with stakeholders, from government officials to women filling their water containers revealed opinions on the roles for themselves and the other stakeholders. The following section describes some of the roles reported for each stakeholder.

10.3.1.1 The village Mayor (or the Chef du Fokontany)

The mayor was the most referenced person as important to be involved in a water project. In Madagascar, each village has a Chef du Fokontany, while a commune consisting of 5–20 villages collectively elect one Mayor. The Chef du Fokontany and Mayors were universally trusted by their constituents, who felt that they did a good job of leading community meetings, listening to the village's opinions and maintaining the trust of their constituents. People reported adopting whatever recommendation or opinion was offered by their mayor. Villagers saw their mayor having the responsibility to uphold their right to water, and this was done through appealing to regional actors for resources, meeting with prospective water partners to screen for useful opportunities, and recommending the most promising safe water ventures. Further, along with the water users, the Chef du Fokontany and mayors were responsible for acting as (or appointing) the 'water project owners', usually the management committee. Finally, the mayor facilitates community meetings, including the introduction of potential water project partners.

10.3.1.2 Government and the Ministry of WASH

In Madagascar, this is organised by a federal ministry and regional directorates. The federal office is tasked with establishing national water, sanitation and hygiene objectives and a plan and budget to achieve each. A programme director from a large private sector actor added that the federal government should foster "*an enabling environment so that all actors will be welcome to the country and to the sector to support achieving that objective.*" This role was reiterated by others too: that the regional- and national-level governments should set objectives for water and health, which are then carried out by local-level programming. And by creating an enabling environment, the government facilitates inviting other stakeholders to the region to carry out these already-established objectives.

In this role, the regional ministry can translate federal objectives into localised regional and village-specific targets, then identify priorities and gaps and establish a budget to meet these objectives at the village level. Furthermore, the regional ministries can maintain relations with villages and water sector actors, and transmit local-level feedback to inform federal objectives. Foremost, the Regional Directorate of WASH was responsible for coordinating activities by all actors in their region. For someone coming with an idea or question, the regional directorate knows the people to approach in a region because they've been coordinating with all the region's actors.

10.3.1.3 Teachers and Nurses

For matters of health, people trust their nurses and doctors. For matters of education, including water and health, people trust their schoolteachers. As such, these two stakeholders have an important role in a water project, especially during the early steps to raise awareness of water contaminants, their connection to family health, and the education of possible solutions. A strategy shared was to ask the local nurses to run a session on explaining transmission routes for faecal contamination. Rather than a foreigner from Canada running this session, instead ask what resources the nursing station might want to be able to host a workshop: posters, plastic toys depicting bacteria, or lunch for the women attending?

10.3.1.4 Donors

In many cases, donations were filling gaps in the government's provision of services, where such gaps were being identified in sectoral reviews by prominent NGOs. Local stakeholders stressed the importance that this support be offered towards *existing domestic priorities* rather than the donors' perceived ambitions. Donors can increase the impact of their support by ensuring that the project's decision-making power remains as close to the end-users as possible. The donors can achieve this through a review of the national or regional plan for progress on water and sanitation or even by approaching a local organisation and *asking* rather than *telling* them what they could use the funding for. It is also important for donors to choose the right local organisation based on their track record, business organisation structure and likelihood of success. One focus group suggested that during funding negotiations, the NGOs or implementing partner may submit a CV highlighting their track record of project delivery, as opposed to solely their qualifications, since formal qualification certificates may vary depending on access to education and may not represent an organisation's potential for changemaking.

10.3.1.5 International NGOs

These organisations should align their priorities with existing domestic objectives. The role of the private sector was largely described as supporting areas where governments seek to provide services but do not have the resources to do so. Large international organisations expressed their role as 'providing financial and technical services' to support both local implementing partners and the volunteer water management committee and the national government alike, with training ranging from developing water safety plans to bookkeeping and financial accountability to templates for hiring contractors.

The private sector also reported having a significant role in capacity building. NGOs have the resources to run leadership workshops, technical training programmes, and water user education sessions. One remark was that "*NGOs need to provide the technical and organisational capacity to the local authorities and communities so that the two can be involved."* Importing foreign expertise, especially through leveraging virtual workshops, can be an exciting way of building capacity. For instance, training in bookkeeping for volunteer water management committees or working with the Ministry of Health to establish surveys of water and sanitation access in their regions. The private sector also expressed a need to improve coordination amongst the water sector actors. One suggestion was to take better advantage of existing events like World Water Day, school openings, etc. to host forums that share successes and foster collaborations.

10.3.1.6   Local NGOs and Implementing Partners

Village-level implementing partners have the tireless role of being the connection between all other stakeholders while possibly also managing the water project delivery. Firstly, since local NGOs typically comprise members of the communities that they serve, they can have the valuable trust of these communities. And so, they have the role of interacting with the communities through a two-way communication of knowledge. They may collect information on the local priorities, values, and history of successes and failures of past water projects, sharing this information with other actors, including the government and international-level stakeholders. They may also be the right stakeholder to carry out awareness and education sessions or to directly support the water management committee during the establishment and then through the operation of the water system. In this case, local NGOs also serve to provide technical services. However, aware of their limited resources, it may be successful for an international NGO to support the local NGO with program materials like technical construction manuals, templates for contracts, and advice for setting a water tariff, for example. Meanwhile, the local NGO can translate this knowledge into the local context and share it with the community. Local NGOs may also more directly support the construction phase of a project and be available for ongoing support during the operations phase, but their role is that of the central body in two-way knowledge-sharing between the water users and programme managers.

Implementing partners likely also have a role in recruiting financing and maintaining donor relations. In this capacity, they should tread the line between appeasing donor intentions and maintaining objectives focused on locally established targets. This could be done by offering donors a selection of the most developed water project proposals so that donors might choose which proposals to support.

10.3.1.7   Private Sector Entrepreneurs

While hosting awareness and education sessions, a local implementing partner may be encouraging households to adopt safe water and sanitation practices like using a household water filter and keeping a clean latrine. In turn, this creates a demand for safe WASH supplies such as replacement ceramic filters, toilet bowl cleaners and handwashing buckets. This market demand creates an opportunity for the private sector to establish a supply chain and earn small profits while providing these materials. A local hardware store, cutlery shop, or other store with existing supply chains for brooms and buckets may be the right partner for this service. There is a delicate balance between social responsibility and ensuring sufficient profits for the hardware store, but government subsidies can help this, as can continued demand generation through awareness campaigns led by the implementing partners.

10.3.1.8 Technical Experts

The role of technical experts is to understand the plethora of water treatment options available and devise alternative solutions to the challenges raised by water teams. Instead of promoting a single water technology, technical experts should advocate for safe water broadly and have a toolbox of options combined with the engineering dexterity to devise and adapt a technology to a local context. This is best done by understanding the underlying performance and functionality of a water technology instead of its form. This knowledge empowers technical experts to work with local artisans to localise safe water technologies with locally available materials and construction techniques.

Technical experts also have a responsibility to build trust with water users. Discussions with non-technical persons revealed a preference for a prototype demonstration and analogies to commonly understood concepts. For example, a solar desalination system might be described as *"mimicking the water cycle by evaporating the salty ocean to make clouds, then collecting the clean rainwater, all within this concrete box."*

A beautiful quote by the Regional Director of Infrastructure encompassed the role of technical experts: *"As an outside stakeholder, you should not seek the problems. The problems are the community's problems. The solutions are also the communities, and they know the solutions that are adapted to their needs. But it's the technique that's missing, and that's what can be contributed by technical experts."*

10.3.1.9 Water users

Of course, the water users themselves are a critical part of the water partnership. Their role is largely to participate: to share opinions, experiences and cultural practices, and to give honest assessments of the proposed solutions. Only water users can situate a proposed solution within their local context, and this responsibility should not be overlooked. Along with the mayor, water users could act as the 'owners' of a water project. Through introspection, the end users can also identify the village's realistic capacity for project management and where trusted organisations can be asked for help. Finally, in many (although not all) households, women are the caretakers of water. As such, their opinions should be shared in water team discussions: what would make them excited about a new water solution? Would a proposal necessitate a behaviour change, and would this be desirable? Such questions and their answers can be informed only by the water users themselves.

10.3.2 Tools to Host a Community Water Meeting

Central to the partnership model is the necessity of inviting opinions from all stakeholders. This study spoke with water team participants about strategies for hosting meetings and respecting everyone's voices. Firstly, the meeting format should change depending on the audience. Meetings with government officials and business leaders may involve scheduling a one-on-one meeting in an office during the workday.

Meanwhile, meetings with water management committees, system operators, or water users may be less formal, involving a phone call the day before to ask for availabilities and then sitting on plastic chairs under the shade of a tree in someone's yard.

Receiving water users' opinions requires a meeting at the village or commune level. The town hall, commune building or an existing water handpump was reported as a good venue for water meetings. However, each community almost certainly has a designated meeting spot, and the mayor will know the proper location. If the mayor is not available to host, then another prominent community member could be recruited. Often, a neighbourhood leader can lead the discussions, invoking a 'leadership from within' approach. However, it can be useful for the meeting host to discuss with the NGO partners before the meeting to determine meeting objectives and perhaps brainstorm some suggested courses of action.

The time of day may influence the audience. In general, women and mothers were reportedly more available for daytime meetings, although men do show up then too. It is also necessary to give at least one week's notice before a meeting. Each rural Malagasy village has a market day once per week, and it is during this event that information gets passed around, including meeting invites. Giving at least one week's notice ensures that a market day happens before the meeting date.

Fairly unanimous was the need to share the invitation with the whole community. Word of mouth was the most important mechanism: market day gossip, chatting at the water point each morning, or even the Chef du Fokontany going door to door to invite villagers. If the project for discussion involves a water point at a school, the parents' organisation could be invited, and the school's head teacher may help to mobilise support.

Typically, in Malagasy culture, decisions are made by consensus, either by voting or by the discussion arriving at a common understanding. The meeting might start with the Chef du Fokontany introducing the problem or a possible solution. Discussion ensues, with villagers voicing concern or support for the idea. The discussion may easily diverge into other areas of concern, yet by the end of the discussion, a consensus emerges. Unlike in Western societies, where people often remain rooted in their opinions (leading to a meeting concluding without a unified decision), traditional societies might adopt a more open mind with the intention of finding common ground. Some differences between traditional and Western societies are described by Jared Diamond in his 2012 book, The World Until Yesterday . Once a decision is adopted, everyone who agrees may sign their name on a piece of paper outlining this decision, indicating their support.

Largely, discussions in Madagascar revealed that women were always invited and were usually equally as present as men at water meetings. However, women were perceived as quieter and less likely to participate compared to the louder and more opinionated men. While women were happy to share their opinions amongst themselves, they reported preferring to support the ideas of their husbands and neighbours in group settings. Their important opinions may be encouraged by hosting women-only breakout groups. For instance, after ten minutes of group discussion, the participants are split into two groups: the men and the women. These groups discuss amongst themselves for another 10–15 minutes, then representatives share ideas back with the larger group. This strategy for breakout groups may equally be applied to divide groups by gender, age, class, occupations or otherwise, to encourage idea sharing amongst people who are comfortable with each other.

10.3.3 SUGGESTIONS FOR SHARING TECHNICAL INFORMATION

As was quoted above, villages know both their water challenges and solutions. However, in some areas, water users may not know what a water or sanitation solution looks like. This is an opportunity for technical experts to fill knowledge gaps or brainstorm options. Non-technical water users expressed a significant preference to be shown a prototype and a simple demonstration instead of a written report or poster. Further, experts can achieve better technical sharing by focusing on the *impacts* of a water technology rather than its *mechanisms*. For instance, a rainwater system can provide water for household use from a tank close to the home and even on days without rain. By describing the time savings as not having to walk to a distant water point, the rainwater system can be described by its impacts: saving time on water collection as well as providing a clean water source. This can be more effective than explaining the technical design of gutters and first-flush systems.

Describing materials and appearance was also important for end user acceptance: a large, blue concrete tank with a lid on it, which will have a small water tap just off the ground with enough space to fit a jerry can underneath, and grey plastic pipes from the roof gutters to collect rainwater into the large concrete tank. Water users are also interested in the costs to them of this system, which is a discussion that should include local leadership. And of course, people want to know if the water being delivered will keep their family healthy.

Once again, sharing technical information was easier when paired with a trusted local technician. Rural villages may look to their local engineer or technician and adopt this person's suggestions. It is important to identify this trusted technician and include them in the water partnership. The water team not only needs their support but also their expertise on what has worked and failed in the past.

The use of analogies to commonly understood concepts ("the concrete basin is a big water bucket to collect rainwater") was suggested by technical experts, and the mantra 'simple and straightforward' was commonly recommended.

It is important to also respect the water users' intelligence while communicating. Although the water team may not have a civil engineering background, they can understand the operating principles of most water technologies and are certainly familiar with water sources and challenges in their villages. For example, one rainwater technical expert explained that people knew about viruses and bacteria in their water, and while they may not know the killing mechanism of chlorine, you don't need to 'dumb down' your technical knowledge dissemination.

10.3.4 TOOLS FOR BUILDING TRUST IN COMMUNITY WATER PROJECTS

Safe drinking water is essential to health and enjoyment of life. It is important to understand that water users are placing trust in a water solution. Why might someone trust a rainwater harvesting system? As discussed in the introduction, trust is a prospective assessment of someone's ability to do good work. Hence, it is important for the water team to work closely with the community and to be transparent in their intentions so that the community can begin to *trust* the members of the water team.

Through discussions with water stakeholders, five categories of strategies were identified to build trust between communities and foreigners. Recall that 'foreigners' can be someone with whom a community doesn't normally believe to share their values and experiences, even if they are of the same nationality.

1. *Adhering to local customs*: In Madagascar, this begins with a formal introduction, which is best left to the experience of older and respected community members. This can show that the partners are approaching the relationship from the community's traditions rather than imposing their foreign customs (remembering that 'foreign' may be of the same nationality). Building trust was reportedly better received when using the local language and dialect. The foreign-born director of an implementing partner explained, "*I operated for years through a translator, but you miss a lot. As you're waiting for the translation, your mind starts wandering and there's a bit of distrust around 'well, what are they actually saying.' But when you speak the language, you can apply charm and you can build a relationship.*"

    Adhering to local customs also involves understanding the relevant local governance systems. This involves the respectful conduct of meeting first with the mayor but also understanding applicable laws and rights. In Madagascar, the Code de l'Eau says that water is universal (LOI nN° 98-029[1]). But while the water belongs to everyone, users must pay for the service of water provision, a policy that is in line with the U.N. resolution recommending that people contribute financially or otherwise for water services. Further, in Madagascar, built water infrastructure belongs to each village, so they must be the ones making the decisions on what is needed, appropriate, and manageable with local resources. Foreign stakeholders should adhere to such local customs when building trusting relationships.

2. *Align the water project with trusted people*: There is often less *confidence* in projects led by foreigners, and for good reason: there are historical patterns (i.e., a track record!) of outsiders coming and going. Inherently, there is greater trust for locals, and so one might find success through a local advocate to introduce and promote the water project within their own community. While these champions may be difficult to find, they are likely also the first ones to show up to a community meeting or to inspect prototypes. In rural Madagascar, aligning with trusted people also means speaking with the Chef du Fokontany, who is well-respected by their village. In some cases, you could align with another organisation working in the area, being certain to set clear roles and responsibilities and ensure complementary visions.

3. *In-person engagement*: This is especially true in communities where internet, email, and virtual relationships are not commonplace, including in rural areas of developing and developed countries alike. Being visually present and attentive to people's opinions, needs, and ideas goes a long way towards building trust.

4. *Long-term relationships*: Amongst water users, one of the most important tools for building trust is long-term engagement. An acquaintance with whom someone shared a few coffees over a few years may seem like an 'old friend' who may be trusted more than a new colleague with whom they have worked full time for the past month, despite knowing more about the newer colleague. And so, it is useful to start introductions early and often. If the timeline doesn't permit years of relationship-building, 'aligning with trusted local partners' can be effective for leveraging long-standing relationships with other organisations.

Engagements to build long-term relationships mustn't be all-business, as explained by the Regional Director of an international NGO: *"Yes, to have discussions with the people, even if they're symbolic, it's important to show interest. For example, you could host a focus group to ask the opinions of the beneficiaries in the community, which shows that you are truly interested in the project."* In fact, this could be organising a football match or participating in a weekly market to show genuine interest in supporting the community and their values. In another discussion, it was suggested that 3–4 visits over a year or more may suffice, but the benchmark is when you understand the local context, and the partners trust each other's intentions.

5. *Respect for local capacity*: This tool highlights the mistrust between on-site and off-site actors. By the end of a project, it is usually clear that there are talented local individuals to drive success, but too often, foreign actors prefer importing talent from elsewhere. Respecting local capacity means ceding project management to local champions: letting the local community establish their values and priorities around water and health, inviting ideation from the community, and having water system designs inspired by local artisans and selected by a process led by local leaders. This not only maintains local ownership but also creates an opportunity for local leaders to show their talents and rise to the occasion.

Other recommendations for building trust include being informal during initial meetings by avoiding the use of cameras, survey equipment and notebooks, working to find common objectives amongst collaborators and not being too pushy in negotiations. It is also useful to briefly recognise three sources of community *dis*-trust, reported also by Enang et al. (2008) on a water project in Kenya. The first is termed *'paying for another man's sins'*, and results in distrust of ongoing initiatives due to a history of failed interventions by past do-gooders. The second is *'insufficient social capital'*, manifesting as a lack of a community's ability to maintain momentum and progress on a project, such as a failure to attract sufficient labour to transport gravel or to dig a pit for a water tank foundation. The final source of distrust was *'mis-information'*, characterised by the spread of fake news and gossip around the intentions or form of the water project.

Such distrust could result from a mindset of 'us-and-them'. The first and third sources may be remedied with greater up-front efforts towards relationship building: being transparent and finding common ground with each party's intentions may help to understand with whom you are partnering. The second source results from a lack of local engagement and ownership. By having a local water champion to maintain engagement and excitement and by ensuring transparency and widespread understanding of the new water project's intentions and benefits, stagnation can be avoided. In all three cases, distrust results from a mindset where a foreign actor is importing a water solution with varied levels of community enthusiasm. Instead, the partnership approach flips this mindset upside down, placing a project's definition and management with the community, while the roles of other stakeholders are to support the community's needs.

10.4 CASE STUDIES IN WATER PARTNERSHIPS

10.4.1 ELEMENTS OF A WATER PARTNERSHIP

For the readers of this book who are educated people in high-income areas, how can we participate in water partnerships under the vision of community ownership? This section will explore a hypothetical case: the town of Riverbend in a temperate climate. Riverbend has about 380 villagers spread out over a kilometre, with some households closer to plantations on the outskirts of town. Through this narrative, a Water Partnership approach will be described with the following steps:

1. Building relationships with the community and existing organisations.
2. Establishing common ambitions with end-users and other stakeholders.
3. Needs and capacity assessment.
4. Consideration of technical solutions and their assessment for the local context.
5. Localisation of technical solutions based on end-user input. Final decisions are made by local leaders.
6. Project design and construction, emphasising capacity building for local tradespeople and artisans.
7. Establishing an ongoing water team for operations and maintenance oversight, fee collection, user assistance and feedback mechanisms.

The first step in a water partnership is to speak with local leadership, the mayor, the city council, or an existing water provider. You may be able to have email contact, but soon enough, an in-person visit is the best for building a relationship. After receiving permission to continue, one may seek to partner with an organisation having long-standing relationships in the area. The organisation may be from Riverbend itself or an actor in that province. During these initial meetings to define the scope and roles of the water partnerships, it is important to understand the local priorities and values around water and health. Does a plan exist for a piped water network? Is there an underserved population for whom the mayor has been trying to get water access? The role of water partners is to support locally led initiatives, so it is important to ask and listen to what these initiatives may be, but don't try to change values or to lead by explaining what you believe that the community needs.

This is also the right time to establish roles and responsibilities so that all partners are aware of expectations and contributions from each other. One of the first tasks is to establish common ambitions. This must consider the user-perceived needs and align the project team towards these ambitions. By explaining your organisation's background and interests, this transparency also lays the foundation for trust.

The next step may be an assessment of the needs, the willingness to participate, and the capacity to participate. This includes listening to the local leadership and speaking with water users directly. Depending on local customs, this may include conducting surveys or eliciting written feedback in a suggestion box, but usually in-person chats are best. Perhaps most effective is to simply spend time walking around the village, listening, and having casual conversations. Chickannaiyappa (1984) described this phase as "*unstructured interviews with community leadership and water users to identify users' attitude and preferences.*" One stakeholder said that their first task was to spend a week in the village having casual conversations without holding a notebook or camera. Simply chatting with people about water. On a stay in Riverbend, you might hear at a Tuesday lunch that the whole village comes to watch a football match on Thursdays. At the match, you notice a large water truck providing water to the players but also a dozen women lined up to fill water buckets. These women live in homes in the highest part of town, and in conversation, they tell you that although Riverbend has a piped water supply, the town's water pumps are not strong enough to force water up the hill to their houses, and so they must walk to the community's water point at the bottom of the hill. This football field is halfway up the hill, making Thursday's water collection a comparable pleasure. This is essential information about the community's water needs and existing water systems to build off.

The next step involves better understanding the village's water needs, existing ideas or preferences, the history of successes and failures, and establishing an under-standing of the short- and long-term commitments of each party. As these discus-sions evolve, it is natural to invite more water stakeholders to the discussions: local technicians, elders, regional or federal governments, donors, etc. In a meeting led by Riverbend's mayor, some people in the community might explain their displeasure with the intermittency of the piped water supply. The mayor agrees but explains that he's more concerned with first ensuring universal access to water, and so for this meeting, he invited the women from the homes on top of the village hill. While most were busy, one woman brought children from three of the families to join and explain their water challenges. Here, the community agrees that the first priority is ensur-ing universal water access, and a consensus emerges to bring water to these hilltop homes. A common ambition is established.

As an understanding of Riverbend's perceived water needs is built, the water team may begin considering technical solutions. The foreign partners could inquire with their networks of water professionals for suggestions to improve water supply for these hilltop homes: bigger pumps, repairing breaks in the water pipes, storage tanks that fill overnight, and imposing load-shedding arrangements to ensure sufficient water pressure. The water team should also invite suggestions from local and nearby water experts. In Riverbend, inviting the Regional Ministry of Infrastructure to an evening meeting brings forth other suggestions: a nearby community tried a solar pump, which worked for a while, but the initial costs are very high. They've also heard of floodwater harvesting, where water is diverted from streams during the rainy season and stored in tanks or ponds for later use. Here, the technical experts' role is to provide a feasibility assessment of technical solutions, their likely impact (including other social, economic, labour, etc. impacts), and costs.

At the next water meeting, the five couples from the target houses attend and offer their opinion that floodwater harvesting is interesting, and in fact, they have an unused 8,000-L concrete tank built by a previous NGO for trucked water deliveries. However, someone else points out that the nearest stream is 2 km away, too far for piping waters. But a suggestion emerges: there's a warehouse nearby – what if rainwater could be collected from its large steel roof? At this point, your foreign connections can once again support this water team: you know a rainwater harvesting expert, a professor in Thailand! You pitch the idea to the Thai professor, and they agree it's feasible and that they could draft a rough plan and send a student to help train the local construction team on a few designs and construction tricks, along with a plan to monitor water quality and user experiences.

After proposing this offer to the town's water committee, they invite the rainwater harvesting proposal, and the student's visit is scheduled. The Regional Minister for Infrastructure offers to provide a loan for water access if the community contributes in kind for half of the costs. At the next Thursday football match, the mayor asks for volunteers for the project. The engineering firm owning the building (and roof!) is also excited to offer their roof, a win-win for their public image.

Months later, villagers have brought gravel and cement. The Thai rainwater expert hosted a video call with the local water committee to explain the details of the rainwater system, including the first-flush diversion, the importance of the cleanliness of the roof, and the suggested maintenance intervals. During construction, it is critical to establish a Water Management Committee; ideally, the Mayor would ask for volunteers, intrinsically giving his support, and especially including the hilltop residents. This water team should address regular maintenance, equitable collection of fees, each home's water withdrawal privileges and occasional water safety monitoring.

This fictitious story of Riverbend highlights the natural feeling of a partnership model: the problem and the solution both emerged from local ideas. Foreign expertise was brought in to support the project as needed, but ideation, management, and project execution were led by community leaders and water users. The solution was not imported from abroad but involved collaboration between the Regional Director, the water authority, a local business, end-users, and the foreign technical expert.

10.4.2 PARTNERSHIPS IN PRACTICE: A CASE STUDY IN RURAL CANADA

Finally, we consider a case study of a successful water partnership to support a rural community in Canada,. The houses in this village are dispersed so that many have single-home water systems. The region's underlying geology meant that some wells had fluoride, iron, and manganese. In addition to hardness, which quickly corrodes plumbing, washing machines and shower heads, iron can stain clothing orange, and fluorosis can emerge with prolonged fluoride exposure, especially in children (although there were no cases of fluorosis reported). The village wanted a solution for contaminant-free water.

Typically, the federal government agency responsible for infrastructure would hire a firm to conduct a feasibility study, then detailed designs, then tender the proposed solution for construction; community administration would likely be consulted and give feedback on designs, but not directly involved in the design decisions that are proposed by the engineering consultant. But in this case, the community asked a partner organisation called RESEAU Water'Net to facilitate a water partnership through their Community Circle™ approach. A water team was assembled to include the local village administration and water operator, the federal government agency (which was funding the project in addition to providing technical support), the health authority, equipment vendors and trusted contractors who were identified by the community. The project began with community consultations, and the local leadership chose two houses for a pilot test before scaling to other households.

Instead of a conventional design-bid process, the group adopted a risk-sharing approach and were comfortable with this because of trust in the water team, who invested a year's time to understand each other's interests, visited the community to learn their needs and values, and generally facilitated an amicable collaboration rather than a business-like relationship. The partnership approach also meant that the water team avoided the expensive and lengthy process of hiring feasibility, design, and construction management contractors. Instead, expertise was leveraged from within the water team. Importantly, all water partners were included in every meeting, and every decision was made by consensus and with input from all stakeholders. But the final decision always rested with the community leadership. In every meeting, the local representatives could share their opinions and concerns, while the equipment vendors and civil works contractors could listen and offer their expert opinions. The health authority gave assessments of what treatment options had proven successful elsewhere, and there was a definitive sense that all stakeholders were on the same team. This way, the group got to understand the users' water needs, information about local water knowledge, and users' perceptions of, for instance, how the iron-contaminated water was staining their skin and clothes, as well as learning from stories of past successes and failures both in this community, and others like it. Throughout, there was a harmonious balance of power, with everyone having equal knowledge in the decision-making process. Rather than *importing* from another village, this solution was *developed in partnership with the water team*.

The solution was agreed to be a point-of-entry treatment system with filters for fluoride, iron, and manganese removal and particulate filters in combination with an ultraviolet disinfection unit. The partners maintained engagement throughout a multi-year monitoring phase, and the homeowners expressed interest in reviewing water quality results too. Here, involving the community leadership and their maintenance team during the design process helped to understand the technical solution and build trust. Meanwhile, end-user engagement throughout the performance monitoring phase established a track record of success, offering the basis for confidence in this solution. With the success of these two pilot homes and the trust and confidence they built, the community was ready to approach solutions for the remaining homes in the village.

Participatory engagement schemes suggest that you *engage* the water users during the project definition and ideation phases before encouraging *participation* in operations and maintenance. While this view appropriately gives water users a voice, we suggest a subtle but significant change: not only engaging the end users but finding a way to support the end users with the resources and knowledge so that *they might engage the rest of the water team for the requisite support*. In the above example, the water team was able to support the community with resources, funding, and expertise as they pioneered their own safe water solution. Yet the partnership approach ensured that local knowledge was valued and that the water users played the role of experts on their village's environment.

10.5   DISCUSSION

10.5.1   A CRITICAL LOOK AT THE PARTNERSHIP MODEL

The Water Partnerships model emerges from roots in participatory processes and end-user-led development. Broadly, this may be more successful in rural communities where the government may not provide their target service levels. For a commodity like water, there are significant cultural, economic, and social values intertwined with its benefits to health and wellbeing. In such situations, it is important to adopt a model that values end-user input, trust and relationship-building. The partnership model strives to:

– Empower end-users with meaningful decision-making powers, *supported by* other partners.
– Include the community's values and priorities in determining a safe water *solution* that considers water's role within a village's specific context, as opposed to merely finding a water *technology* that solves a specific technical challenge.
– Unite locally available expertise and materials to adapt or develop a solution that can be sustainable for long-term water access.
– Build capacity within the local community, ideally permeating beyond the water solution.
– Leaving project ownership and management with the end-users, maintaining passion to operate and maintain the water solution with minimal outside assistance.

There are, however, situations where a partnership model may not be appropriate. Large cities may have sufficient resources for maintaining large-scale water and sanitation infrastructure. In such situations, residents may prefer that their water tariffs (or tax dollars) be used by a utility provider without involvement in each nuanced decision. Such is the case in many modern Western contexts. However, as Western cities move towards new water management approaches like reuse and resource recovery, end-user involvement may again become necessary for public acceptability or preventing challenging contamination in reuse schemes, for example.

Alternatively, there may be communities where, even if participation is desired, local management capacity does not yet exist. In these situations, there may be a good reason for a government to share at least partial ownership with the local management committee, especially if the water intervention is part of a regionally coordinated effort that might offer support. Foremost, there may be local-level capacity constraints. The partnership model assumes local capacity for managing the water project: not only the requisite time but also a level of expertise and connections with businesses, private sector actors, and the government, as well as the ability of some organisations to mobilise funding and technical expertise. Cleaver et al. (2001) summarise this when they state that *"there is a myth that 'communities are capable of anything'."* Indeed, there exist villages where these requirements fail to be satisfied. Such cases may emerge, especially in informal settlements or camps of displaced persons. Villages without the necessary capacity may benefit from a trusted partners helping to determine where best to divest certain management aspects to another trusted organisation.

Finally, the partnership model is slow. Many funding agencies want a final report 6, 12 or 18 months after releasing the budget. But simply building a trusting relationship could take this long, let alone completing a water project through a partnership approach. However, by investing more time up front, water teams earn a generous multiplier on the likelihood of a long-term success for their safe water solution.

10.5.2   What is the Role of Foreign Actors?

Readers of this chapter, like its author, may be residents of a developed country hoping to do good in the most useful manner. So, what type of assistance is most useful? First, build relationships. The chances of making change in the world are much greater by partnering with an established organisation than by forging new organisational roots from thousands of kilometres away. There are lots of deserving organisations out there, and there's a great benefit and pride to come from contributing your resources to support them. If one is privileged to be involved with on-the-ground efforts, adopt the mindset that the village and the end-users are the 'owners' of their project and that the other stakeholders are there to support them. This framing goes a long way towards improving respect. In doing so, make sure to align 'on the same team' as the water users, not as a client-owner relationship.

Foreign participants can think about how to build a trusting partnership rather than beginning with devising a technical solution. What can be done to understand the local context? What knowledge and capacity are available locally, and how can they be promoted and integrated? What gaps in water and sanitation solutions need filling, are there existing plans underway, and what can be accomplished by the existing local capacity versus can be developed into it during the project versus needs filling by outside expertise?

Considering these types of questions from the outset of the water project allows for inviting meaningful contributions from all water partners. To achieve this, foreign actors must trust in the capacity of the local implementing team and community leaders and instead adopt the role of providing 'help-when-needed.'

10.6 CONCLUSIONS

The partnership model represents a new paradigm in sustainable development. It's about pivoting from the 'experts know best' mindset of Descartes and from the idea that 'the privileged must help the under-privileged' that permeated the late 20th-century import model of development theory. Instead, we may adopt a mindset focused on building partnerships to enable end-users to identify their own water challenges and their own water solutions and to build capacity to enable those solutions to fruition. Meanwhile, other water partners can support local water leaders with technical services, financing options, and management support. But importantly, the decision-making authority must remain with local leadership.

The development sector is embracing the need for water partnerships, but there is not yet a cohesive view of what this looks like, and there may not be one. Contrary to past approaches where funding agencies, technical experts, and NGOs had defined roles in the process of importing water technologies, the sector is learning that sustainability comes instead from a partnership approach with ill-defined roles. We must nonetheless remain rooted in the essence of a Water Partnership: to remember that we are on the same team, the goal is safe water for everyone, and that we must support everyone to contribute their strengths. By positioning as a partner alongside the community and the local organisations, the next time that you find yourself waiting for a ferry ride at 10 pm in rural Madagascar, at least you'll know that you are surrounded by a group of friends with whom to share a bottle of rum and to laugh as you struggle to explain a card game in a language where you can barely count to ten.

NOTE

1 Assemblée National de Madagascar (1998), LOI nN° 98-029 Portant Code de l'Eau. Accessed December 2023 at: https://www.assemblee-nationale.mg/wp-content/uploads/2020/11/Loi-n%C2%B0-98-029-Portant-Code-de-l%E2%80%99Eau.pdf

REFERENCES

Arnstein, S. (1969) A ladder of citizen participation. *Journal of the American Institute of Planners*, 35(4), 216–224, DOI: 10.1080/01944366908977225
Assemblée National de Madagascar (1998). LOI N° 98-029 Portant Code de l'Eau. Accessed December 2023 at: https://www.assemblee-nationale.mg/wp-content/uploads/2020/11/Loi-n%C2%B0-98-029-Portant-Code-de-l%E2%80%99Eau.pdf
Chickannaiyappa, M. & Ramaswamy, M.V. (1984): "Community participation - water supply systems in Karnataka". Loughborough University. Conference contribution. https://hdl.handle.net/2134/29079
Cleaver, F. (2001). "Institutions, agency and the limitations of participatory approaches to development". Edited by: Cooke, B. and Kothari, U. 36–55. London: Zed Books.
Collier, P. (2007). *"The bottom billion: why the poorest countries are failing and what can be done about it."* Oxford University Press, 0199740941, 9780199740949
Cornia, A., Jolly, R. & Stewart, F. (1987), *Adjustment with a human face*. Oxford University Press, for UNICEF.
Cornwall, A. (2006), "Historical perspectives on participation in development". *Commonwealth and Comparative Politics*, 44(1), 62–83, DOI: 10.1080/14662040600624460

Diamond, J., (2012), "The world until yesterday: what can we learn from traditional societies". Penguin Books.

Enang, R., Ikpeh, I. J., Reed, B., & Smout, I., (2018), *"Community confidence in WASH and development service delivery: a case from Kianjai, Kenya"*. Loughborough University. Conference contribution. https://hdl.handle.net/2134/35648

Lugard, F. (1922). The dual mandate in British Tropical Africa. Blackwood & Sons.

Onda, K., LoBuglio, J., and Bartram, J. (2012). Global access to safe water: accounting for water quality and the resulting impact on MDG progress. *International Journal of Environmental Research Public Health*, 9, 880–894, DOI: 10.3390/ijerph9030880

Paul, S. (1987), "Community participation in development projects: the World Bank experience". World Bank Discussion Papers; no. WDP 6 Washington, D.C.: World Bank Group. https://documents.worldbank.org/curated/en/850911468766244486/Community-participation-in-development-projects-the-World-Bank-experience

U.N. Economic and Social Council (1956), *Twentieth Report to ECOSOC of the UN Administrative Committee on Coordination*, E/2931, UN, New York.

UNESCO World Water Assessment Programme (2023), *Water development Report 2023: Partnerships and Cooperation for Water*, pp. 189, ISBN: 978-92-3-100576-3

U.N.-Water Decade Programme (2010), *The Human Right to Water and Sanitation, media brief,* Zaragoza, Spain.

US Agency for International Development (1970), *Increasing Participation in Development: Primer on Title IX,* Section 281, Washington, USA.

von Hippel, E. (2005). "Democratizing innovation: The evolving phenomenon of user innovation". *Journal für Betriebswirtschaft*, 55, 63–78, DOI: 10.1007/s11301-004-0002-8

FURTHER READING

Smith, J. (2008). "A critical appreciation of the "bottom-up" approach to sustainable water management: embracing complexity rather than desirability", *Local Environment*, 13(4), 353–366, DOI: 10.1080/13549830701803323

Weststrate, J., Dijkstra, G., Eshuis, J. et al. (2019). "The sustainable development goal on water and sanitation: learning from the millennium development goals". *Social Indicators Research* 143, 795–810 DOI: 10.1007/s11205-018-1965-5

11 Rainwater Harvesting and Community Water Security in South-west Uganda

Francesca O'Hanlon and David Morgan

11.1 INTRODUCTION

At present, nearly half of the world's population without access to safe drinking water lives in sub-Saharan Africa's least developed countries (LDCs) (UNICEF, 2020). In Uganda, for instance, only 39% of the population have 'effective' access (Water.org, 2018). Much of this demographic does not suffer from physical water scarcity but instead from a lack of reliable infrastructure to provide water access (Qadir et al., 2007). Municipal water systems often do not reach the poorest, most remote communities in low-income countries. As a result, eight out of ten people who don't have access to safe drinking water live in rural areas (UNICEF, 2020). Where municipal systems do reach urban populations, quality and quantity are often inadequate due to poor system operation and maintenance. This has resulted in a lack of trust in centralised water services (SSWM, 2018).

Decentralised water services, such as rainwater harvesting (RWH), are often used to 'plug the gap' and can supplement intermittent centralised services or provide essential water where centralised supply does not reach. Staddon et al. (2018) estimate that 86% of Uganda's population rely on either 'improved' or 'unimproved' decentralised water sources, including RWH, hand-dug shallow wells, groundwater collection schemes and local surface water sources. In comparison with these alternative decentralised water sources, RWH has the advantage of supplying water directly to homeowners, with systems often constructed on household roofs.

In recent years, scholars have developed water security frameworks to assess the relationship between water services, human wellbeing and socioeconomic development. Assessing water services through the lens of water security allows for the less tangible impacts of water access to be identified, helping water practitioners to move away from a narrow 'supply-and-demand' view of water access (Jepson et al., 2017). Water security has become a widely accepted term that communicates the broader social, political, and environmental benefits of water-related services. Rather than focusing solely on the materiality of water access in itself, water security frameworks have been used to assess the less tangible outcomes of decentralised water access.

DOI: 10.1201/9781032638102-19

For this project, a new water security framework was designed through consultation with water practitioners who have expertise in the delivery and assessment of decentralised, community-scale water services.

11.2 UNDERSTANDING WATER SECURITY

Cook and Bakker (2012) published what they believed to be the first review of its kind – a critical analysis of the differences and commonalities in approaches to water security across academic disciplines. Their review highlighted that the concept of water security had become increasingly popular over the previous two decades. Ten years later (at the time of writing), there is still no clear consensus on the definition of water security. Definitions of water security remain diverse. While early definitions focused solely on water quantity and availability, more recent descriptions include a multitude of factors ranging from ecological considerations to governance, water quality, human health and climate limitations (Gunda et al., 2019). Scholars like Abedin et al. (2013) refer to water security as a status, whilst others refer to it as a goal (Global Water Partnership, 2000). In recent years, however, water security has been viewed as a modern framing to help identify the social, environmental and technical challenges associated with water provision (Vörösmarty et al., 2010). There is agreement among scholars that, broadly, framings of water security encourage the use of water to increase economic welfare, enhance social equality, move towards long-term sustainability and reduce water-related risks (Hoekstra et al., 2018).

11.3 RAINWATER HARVESTING AND WATER SECURITY

Scholars have begun to use the lens of water security to assess whether rainwater harvesting has the potential to provide adequate water of good quality to satisfy user demand. Much of this research, though, has adopted a narrow definition of water security as simply *'the percentage of time a tank of a given size is empty'* (Haque et al., 2016; Kisakye & Van der Bruggen, 2018). This is of value when designing and optimising RWH systems. However, further research is needed to better understand the environmental, socioeconomic, and geophysical conditions under which the technology is most likely to be adopted, maintained and of value to the user.

It is critical to distinguish that the rainwater harvesting literature that is of interest to this study is that which places RWH in the context of water security, climate change adaptation and socioeconomic benefit to the user. There is already a robust understanding of the optimal design features of rainwater harvesting (Pacey & Cullis, 1986; Thomas, 2000; Andersson et al., 2009; Mohammad et al., 2017). Much research has been carried out on how best to design rainwater harvesting systems to meet demand based on historic rainfall data (Rahman, 2017; Zhang et al., 2018) and on optimising the tank storage volume so that it is compatible with the size of the rooftop in the system (Liaw & Tsai, 2004; Mishra et al., 2009).

Within this literature review, only five papers have been found that refer to RWH in the context of water security (Bitterman et al., 2016; Kahinda et al., 2010; Kisakye & Van der Bruggen, 2018; Musayev et al., 2018; Staddon et al., 2018). The few studies that do refer to rainwater harvesting in this context fail to describe the multiple characteristics associated with water security.

Musayev et al. (2018) do not define water security at all but find that domestic rainwater harvesting has the potential to improve household water security under several future climate scenarios. In the case of their research, the term 'water security' refers to ensuring current and future access to water. To be effective at mitigating drought, they find that tank size must be suitably large to store enough water to meet predicted demand over the dry season. There are challenges with the practicalities of ensuring a large enough tank size, as cost and availability of space are limiting factors.

Kisakye and Van der Bruggen (2018) also assess the effects of climate change on water security, but critically, reduce the definition of water security down to '*the number of days a tank of particular size is empty as a ratio of the total number of days in a given period*'. Kahinda et al. (2010, p. 744) assess domestic rainwater harvesting in South Africa as a climate change adaptation strategy but define water security simply as the '*percentage of household demand that is satisfied*'.

Bitterman et al. (2016) develop a model of water security from rainwater harvesting for agricultural use in the Tamil Nadu region of India, elaborating further by defining water security as '*the sufficient availability and equitable access to water as an input to agricultural production and associated human wellbeing*'. This definition integrates the availability and accessibility components of water security but does not take a more holistic view of the term, with no mention of the role of technology management, governance or human rights in contributing to water security. In most of these pivotal papers that unite RWH and water security, there is no exploration of the meaning of water security (as described by leading water institutions and academics).

The exception is a publication from Staddon et al. (2018) that describes a 'hydro-social' model of defining water security in relation to RWH adoption trends in Uganda. In emphasising the cultural, social and political relations behind securing water, Staddon et al. (2018, p. 1119) find that '*achieving water security is a dynamic process, bridging the gap between socio-political relations on the one hand and the acquisition of physical water on the other*'.

Very few researchers have bridged the gap between the conceptualisation of water security and the assessment of RWH to improve water security. Instead, the focus has been on assessing whether RWH can physically provide enough water for users. As a result, there is a lack of knowledge of how RWH can meet the range of water security goals that are associated with human wellbeing and socioeconomic development.

11.4 RESEARCH METHODOLOGY

11.4.1 DEVELOPMENT OF A WATER SECURITY FRAMEWORK

In the absence of existing frameworks designed for the assessment of the contribution of decentralised water services to community water security in low-income settings, a new framework was developed for this study. Semi-structured interviews were carried out with stakeholders invested in the provision of community water services in Uganda, where the uptake of rainwater harvesting is below targets set by the United Nations. In some cases, interviewees drew on their experience working more broadly in low-income settings.

Stakeholders included WASH practitioners from global NGOs, water experts from think tanks and government departments, plumbers, water specialists and water operators from Uganda. The intention was to engage both stakeholders with pre-existing knowledge of water services in low-income settings and experts in the field of decentralised water provision in Uganda.

11.4.2 STAKEHOLDER INTERVIEWS

Thirty-two semi-structured interviews were carried out between October 2018 and July 2019. Interviewees were selected based on their organisation's role in the water and sanitation sector, their expertise in WASH project implementation in Uganda, and their willingness to participate. Interviews were conducted face-to-face, via telephone, or via video conference. Initially, ten interviews took place in person during a scoping study in Uganda. Four interviewees were from a local Ugandan university, two were from the Ugandan government, one was from a local plumbers' group and three were from a local Ugandan community women's group. In addition, 48 organisations from the water and sanitation, global development and humanitarian sectors were contacted for interviews. Of the 48 organisations contacted via email with a request for an interview, 24 stakeholders agreed to be interviewed. In total, WASH practitioners, programme coordinators, and monitoring and evaluation specialists from twenty-four INGOs were interviewed. The complete list of interviewees can be found in Table 10.1. An interviewee code has been used to assign quotes to each interviewee.

Interviewees were provided with information about the research project, provided with a participant information sheet and consent form, and informed that their anonymity would be preserved. All interviews were recorded with a digital voice recorder and later transcribed. Interview codes in the form of 'ST#' are used in this chapter to indicate reference to interviewees' comments and to protect interviewees' anonymity.

A semi-structured interview approach, where the interviewer has a series of questions in a general form known as an interview guide, was adopted for the stakeholder interviews (Bryman, 2008). This approach was deemed appropriate as it allowed for both flexibility and a wide range of topics to be covered.

The interviews contained questions on the following four main themes:

    I. Water security meaning and metrics
    II. Current methods of assessment of water-provisioning and WASH projects
    III. Decentralised water services (including RWH)
    IV. The role of water security in water project assessment and implementation.

The complete stakeholder interview guide can be found in the <Appendix/ box at the end of the chapter>. Saturation was thought to have been achieved through the interviews when no new characteristics of water security were identified and when there were recurrent descriptions of the same water security goals.

11.4.3 INTERVIEW DATA ANALYSIS

Interview data was transcribed and then coded using the computer software programme Nvivo 12.6.0 (QSR International). Codes served as devices to label, separate, compile and organise data (Charmaz, 2006). The process of coding was iterative,

TABLE 11.1
Role and Organisation of Interviewees for Water Security Framework

Role	Organisation	Interviewee Code
WASH Advisor	Tearfund	ST01
Rural Sociologist and Gender Expert	International Water Management Institute (IWMI)	ST02
Chief of WASH	UNICEF	ST03
WASH analyst	Department for International Development (DfID)	ST04
Founder/WASH specialist	Centre for Humanitarian Change	ST05
Monitoring and Evaluation Advisor	Save the Children	ST06
WASH Advisor	Independent Consultant	ST07
Sustainability Officer	Life Water	ST08
WASH Coordinator	Impact Water	ST09
Research Fellow	Overseas Development Institute (ODI)	ST10
WASH Engineering Advisor	Concern Worldwide	ST11
WASH Advisor	Centre for Affordable Water & Sanitation Technology (CAWST)	ST12
WASH Programme Advisor	Water Aid	ST13
Programme Coordinator	UNHCR	ST14
Programme Coordinator	The Johanniter	ST15
Programme Coordinator	Just a Drop	ST16
WASH Programme Manager	SEED International	ST17
Operations Manager	eWaterPay	ST18
WASH Programme Specialist	Oxfam	ST19
WASH Advisor	World Health Organisation (WHO)	ST20
WASH Coordinator	Water Aid Uganda	ST21
Emergency Water & Sanitation Engineer	Médecins Sans Frontières (MSF)	ST22
Water Resources Manager	Uganda Ministry of Water and Environment	ST23
Country Coordinator	Impact Water Uganda	ST24
Treatment plant manager	National Water and Sewerage Corporation	ST25
Head of WASH	KDWSP	ST26
Coordinator	Mbarara Plumbers' Association	ST27
CEO	Afrinspire	ST28
Coordinator	Mbarara Development Studies Centre	ST29
Group Leader	Mbarara Women's Group Leader	ST30
Gender Studies Academic	Mbarara University of Science and Technology	ST31
University Tech Hub Coordinator	Mbarara University of Science and Technology	ST32

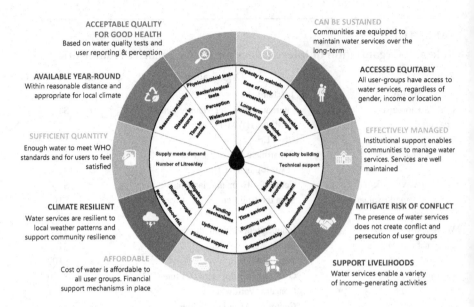

FIGURE 11.1   Water Security Framework developed for this study (O'Hanlon, 2021).

where the researcher moved back and forth between the various stages of coding and the analysis of the data. Throughout the interviews, memo-taking was carried out to assist in the coding process (Bryman, 2008).

The first cycle involved line-by-line coding of the interview transcripts. This resulted in 336 coded segments. This round of coding involved selecting segments of statements and categorising them into groups. The next round of thematic coding aimed to further group and reduce the number of codes, as well as explore the relationship between the codes and then group the codes into wider common themes. Finally, the water security goals were selected from the codes (see Figure 11.1).

Criteria to select the goals for the framework were:

1. Their applicability to decentralised water services on a community scale.
2. The frequency with which they were mentioned by stakeholders. Concepts that had less than ten coded segments attached to them were not considered. This was to ensure both that detailed descriptions of concepts could be provided and that each concept was deemed significant to achieving community water security.
3. Applicability to decentralised water services in low-income settings.

Following the identification of the goals, metrics were developed. During the interviews, questions were asked about current metrics and methods of assessment of WASH projects (see Appendix/box for details). In the initial round of coding, where 336 coded segments were identified, any segment that referred to the measurement of water security or the outcomes of water supply services was assigned the code 'metric'. This selection process resulted in the development of the 32 metrics presented in the framework.

11.5 FINDINGS: THE WATER SECURITY FRAMEWORK

In Figure 11.1, the ten water security goals identified through the stakeholder inter-views are presented in the outer ring of the framework. Visually, the framework can be broken into ten 'slices'. The framework is designed to link theoretical concepts to techniques to assess these concepts. In the central ring of the framework are the metrics used to indicate whether each goal is being met. The outer section of the slice represents the goal, and the inner section includes the metrics that can be adopted when using the framework for post-project assessment.

The framework includes ten water security goals, helping practitioners identify what they should be looking to achieve when delivering water services for better community water security. The structure of the framework encourages researchers to collect a range of qualitative and quantitative data. The framework also provides the basis for post-project assessment of whether decentralised water services have been delivered in a way that ensures community water security goals can be met. In this chapter, the framework has been designed for use to assess rainwater harvesting proj-ects in Uganda; however, it could be used for the assessment of the socioeconomic outcomes of any decentralised water service in low-income settings.

11.5.1 WATER SECURITY GOALS

The identification of the water security goals presented in Figure 11.1 helps to make up a new working definition of water security adopted for this research project as:

> Water services that contribute to community water security provide sufficient water of acceptable quality for good health, which is affordable and available year-round. They sustain livelihoods and can be equitably accessed across all user-groups. These water services should minimise the risk of local conflict and boost community cohe-sion and climate resilience. The management of these services should be supported by local and national institutions so they can be reliably sustained over the long-term (O'Hanlon, 2021).

To demonstrate the rationale behind this definition, exemplary quotations are pre-sented below to justify, explain and provide detail and context for the selection of the goals in the framework. Complete quotations are presented where they provide an illustration of a concept, and interview codes are provided when either more than one interviewee references a similar point, or a description of themes provides a better illustration than a direct quotation.

The wording associated with acceptable quality for good health is important. 'Acceptable to whom?' was a question that several interviewees asked (ST02, ST03, ST13, ST26). Users' perception of water quality was deemed to be as important as physiochemical and microbiological test results. Good-quality water is not a goal in and of itself. Instead, the value of acceptable water quality lies in the reduction in waterborne disease within a community (ST01a, ST06c). Hence, the wording of this characteristic includes references to quality, health and acceptability.

Affordability of treatment was viewed as a significant factor that influenced the adoption of water treatment technologies; however, when water quality was dis-cussed in relation to the other two 'pillars' (ST13) of water security (availability and quantity), it was deemed to be a lesser priority than proximity and quantity of water:

I think quality is valued when quantity is adequate, and I think people do understand very well the links between poor water quality and disease. I think where people can get affordable treatment methods, they will use them, but where they have to pay a lot of money for them, they tend not to (ST09).

Year-round availability was frequently referenced and referred to both the proximity of water sources and availability throughout seasonal variations in rainfall, groundwater and surface water supply. In the context of year-round availability of water in sub-Saharan Africa, extreme weather patterns were viewed as a barrier to the availability of water (ST02, ST04, ST26, ST30): *If intense rainfall events happen at the wrong time, they can massively undermine agriculture. At the other extreme, in arid and semi-arid lands, a lot of the water supply is still surface water. The minute you get high temperatures, you lose those water resources much faster and much earlier in the dry season* (ST02).

Out of the three physical descriptors (quantity, quality and availability) of water security, sufficient quantity was deemed to be the most essential component of a water-secure community. 'Enough' water is measured by the WHO standards of 20 L/person per day and by whether supply of water can meet demand (WHO, 2012). Twenty litres/person/day is deemed adequate for basic hygiene needs and food washing. However, it is still far below the optimal quantity for domestic use, which is 100 L/person/day (WHO, 2012).

It was emphasised that water services needed to be both climate-resilient and delivered in a way that encourages human resilience to extreme weather events and unpredictability. This was cited as a significant challenge (ST16, ST22, ST24). While a number of interviewees (ST03, ST06) questioned whether climate change was a real driver of water security challenges or whether, in fact, long-term interannual variations in weather patterns had always been a challenge for the management of water, several interviewees felt confident that climate change was creating uncertainty around the availability of water resources: *if every year you could predict which months were going to be dry, when the rain was going to come back, and therefore exactly where the ground table was throughout the year, then maybe education around storage and restricting use, or whatever, would get you somewhere. Now, though, things are so unpredictable that no amount of education can prepare you for zero rainfall* (ST07).

Adaptation to the impacts of a changing climate on water access was deemed essential to protecting livelihoods, but interviewees had varying views on what adaptation entails. Some interviewees saw adaptation as retrofitting existing infrastructure: *we've raised the elevation of certain handpumps we installed 5–10 years ago in order to keep them away from flooding* (ST19), while others saw socioeconomic changes such as rural to urban migration as a way of creating jobs that were less reliant on unpredictable weather patterns compared to agricultural professions.

The affordability of decentralised water services was referred to as critical for sustained and equitable access. Stakeholders in Uganda cited the high costs associated with centralised water as one of the biggest barriers to household water access, explaining that despite the presence of water infrastructure, many urban households chose not to connect to the water grid because they could not afford the monthly bills (ST25, ST27, ST30). Equally, the high capital costs associated with private boreholes and rainwater harvesting were seen as a deterrent to the adoption of decentralised water supply (ST26, ST29, ST30). The affordability of water services was referred to as integral to ensuring water access in Uganda: *'In most of the cases we work with, the reason communities can't access water is economic and not due to physical access issues'* (ST23).

Links between funding and whether decentralised water services can be sustained over the long-term were made, with one interviewee stating that *funding is the biggest driver of the sustainable upkeep of [water] services* (ST04).

Most interviewees used the term 'sustainability' to refer to the long-term use of water services, which was cited as a desirable goal by multiple stakeholders (ST09, ST12, ST16, ST17, ST19, ST20). One interviewee believed that the sustainable upkeep of services was the biggest challenge facing water projects in sub-Saharan Africa (ST10d). Whether services can be sustained in the long-term was seen to be decided by both the quality of the projects delivered and the capacity of the community to maintain those projects: *there is a focus on ensuring the hardware is up to scratch, but we need to shift that focus to now ensure water [provisioning] systems can be sustained in the long-term. This is done by ensuring water committees have the capacity and support to carry out maintenance* (ST01).

Water services that encourage equitable access are specifically designed or placed with vulnerable groups in mind. There were contrasting opinions on how this could be achieved, with stakeholders belonging to two camps – those that believed community services encouraged more equitable access and those that believed that household services ensured specific users were provided with water. Vulnerable groups identified included women, children and the poorest users (ST22, ST26).

One approach that has been adopted by stakeholders is to engage women in community water committees (ST19). However, another interviewee cautioned that to improve the quality of life for women, the focus should be on ensuring the proximity of water services rather than on providing women with greater responsibility within the community. This contrast in opinions highlights how equitable access to water services is a contested concept. Questions must be asked about which user groups should be prioritised and how best to engage these groups in the delivery of water services so that a balance can be found between ensuring a sense of ownership and encouraging access for groups that cannot afford household services (ST13).

Effective management with support from institutions was highlighted as a critical characteristic for community water security. The role of institutions in the sustainability and effective management of water services was highlighted by several interviewees (ST04, ST10, ST23). Interviewees' opinions on whether decentralised services should be managed by end users or institutions differed, but engagement with national governments was seen as essential for effectively managing water services: *there has to be some kind of overarching structure provided by governments to support decentralised water services. Day-to-day management may be done by NGOs and community members, but there has to be some degree of support and backstop from the public sector* (ST03).

Mitigating the risk of conflict and encouraging community cohesion were identified as top priorities for two stakeholders (ST19, ST22). When resources are limited, the provision of decentralised water services can create conflict (ST15). Conflict over management of services is common: *we must be really careful not to enter a community and create conflict by exacerbating local tensions over power. In some cases, community services just aren't appropriate, so we provide household solutions, but even these can cause jealousy and resentments* (ST16).

Local 'water gangs' can hijack water services, giving priority to certain user groups and disenfranchising others (ST19). In attempts to overcome this, water committees are set up to manage water services, but it is often members of these committees that are underhandedly controlling and charging for water services (ST22). Embedding a sense of community ownership of services was cited as a strategy to mitigate conflict within communities (ST01).

Finally, the role of water in supporting livelihoods has been well documented as a feature of water security. In some parts of Uganda, women spend as much as 3 hours a day fetching water (ST31), limiting the amount of time they can dedicate to other tasks, including income-generating activities. Reducing the time spent fetching water isn't the only economic benefit of improved water access. To support livelihoods, there must be enough water for a variety of economic activities: *'an absence of water can stop the local economy in its tracks. I've seen construction projects halted half-way because there's not enough water to make the concrete. The projects are just left in that way, sometimes for years'* (ST26).

11.6 DISCUSSION

By 2050, it is estimated that more than half of the world's population will be living in water-stressed regions (Boretti & Rosa, 2019). Achieving clean water access will therefore be at the forefront of the international agenda in the coming decades. Strategies to ensure this access lie in a combination of the technical, socioeconomic and political and can be implemented at all scales, from setting globally recognised targets such as Sustainable Development Goal 6 to ensuring access to affordable household water treatment technologies. Water security provides a framework to develop these strategies and to assess whether they are working. It allows for the complex nature of the relationship between humans, water and the environment to be better understood. It encourages clean water access to be prioritised, ensuring that the current and future water needs of all communities can be met, even when faced with substantial risks such as climate change and demographic transition.

There is still little academic research, however, that links the implementation of decentralised water services to community water security. In particular, most existing literature that links rainwater harvesting to water security principally looks at whether RWH can supply enough water to meet demand, overlooking the impacts of RWH on human wellbeing and socioeconomic development. Water scholars are now looking to understand the wider water-society relationship and the less tangible impacts of decentralised water access. The water security framework developed through this research project aims to provide structure to identify and assess these impacts. The framework is designed to enable data collection in the field, which is useful for researchers, WASH NGOs, water service providers and governmental water organisations.

Some scholars have previously had reservations about the 'indicator' approach to reducing water security to a set of numbers (Jepson et al., 2017). In order to overcome this, the framework was designed not purely to quantify water security but also to define and describe water security through the use of goals and metrics, some of which were quantifiable and others that required qualitative descriptors. The benefit of providing a tool for *assessment* rather than *measurement* is that it allows for the richness and nuance of water security to be captured and ensures meeting water security goals is not reduced to simply 'a tick-box' exercise.

One of the core benefits of the framework approach is that it provides an aggregated picture of water security. The goals allow for the assessment of what is deemed important in the human relationship with water, and the framework provides a wider narrative of what water-provisioning programmes should be trying to accomplish. Jepson et al. (2017) assert that the incentive behind framings of water security is to move away from purely the materiality of water access in itself. The framework developed in this study allows users to assess community water supply in a holistic manner, emphasising the second-order benefits of water access. This provides a method to demonstrate just how critical good water service delivery is to achieving Sustainable Development Goal 6.

11.7 LIMITATIONS

The stakeholder interviews lasted between 30 and 50 minutes. Certain stakeholders were more forthcoming, and so the input of some stakeholders over others was greater.

The selection of interviewees was dependent on their willingness to participate. For example, an interview had been arranged with a senior decision-maker at the Ministry of Health in Uganda but was cancelled at the last minute, so the voice of a significant actor was not represented. Nonetheless, the range of stakeholders involved in the interview series covered notable and significant actors from the WASH and global development sectors.

A notable drawback of the semi-structured interview approach was that not all interviewees were asked exactly the same questions. This can create difficulties in identifying common themes and guaranteeing a consistent research process. However, through the transcription and the rigorous coding process, commonalities in responses and dominant themes were readily identified.

11.8 RECOMMENDATIONS FOR FURTHER WORK

This study has focused specifically on one decentralised water service – rainwater harvesting. Further research could assess the contribution to water security of alternative decentralised water services, including hand-dug shallow wells, groundwater collection services and gravity-flow schemes prevalent across sub-Saharan Africa. This would provide further knowledge on the contribution of decentralised water services to water security and could help with deciding on the most appropriate water service for the arid and semi-arid conditions found in several sub-Saharan African nations.

For this project, the interrelationships between the goals of water security were not discussed. There is a need to further understand how the various characteristics of a water-secure community influence and impact each other so that stakeholders can identify where to prioritise funding and efforts to improve water security. Systems dynamics modelling such as that used by Bitterman et al. (2016) would allow for these interrelations to be further understood.

11.9   CONCLUSION

The aim of this project was to develop a tool for water practitioners, researchers and stakeholders involved in the delivery of water services to ensure that social, environmental, economic, and technical goals are prioritised. The water security framework developed through this research project provides structure to identify and assess the impacts of RWH (and other decentralised water services) on human wellbeing and socioeconomic development. The water security framing presented here can help to structure the identification and assessment of the less tangible impacts of water access, allowing organisations to identify the factors they should prioritise when delivering decentralised water services.

A new definition was developed that describes how decentralised water services can contribute to community water security. They should provide sufficient water of acceptable quality that is affordable and available year-round. They sustain livelihoods and can be equitably accessed across all user-groups. These water services should minimise the risk of local conflict and boost community cohesion and climate resilience. The management of these services should be supported by local and national institutions so they can be reliably sustained over the long-term.

By including goals and metrics with which to assess water security, the framework provided a new structure, not only to define water security, but also to guide the identification of the outcomes of water access. Unlike previous research, the framework developed here puts the sociotechnical outcomes of water access at the forefront of the relationship between RWH and water security. This approach allows for (1) a better understanding of how community water needs can be satisfied and sustained by rainwater harvesting and (2) a clearer idea of the positive and negative outcomes of rainwater harvesting projects and how these can be managed. This research has provided an example of how approaches to water access can be reframed, so that water is rightly viewed as the lynchpin of human wellbeing and sustainable development.

REFERENCES

Abedin, A., Habiba, U. & Shaw, R. (2013) Water Insecurity: A Social Dilemma. First. Howard House, Wagon Lane, Bingley, UK, Emerald Group Publishing Limited.

Andersson, J.C.M., Zehnder, A.J.B., Jewitt, G.P.W. & Yang, H. (2009) Water availability, demand and reliability of in situ water harvesting in smallholder rain-fed agriculture in the Thukela River Basin, South Africa. *Hydrology and Earth System Sciences* 13 (12), 2329–2347. https://doi.org/10.5194/hess-13-2329-2009.

Bitterman, P., Tate, E., Van Meter, K.J. & Basu, N.B. (2016) Water security and rainwater harvesting: A conceptual framework and candidate indicators. *Applied Geography* 76, 75–84. https://doi.org/10.1016/j.apgeog.2016.09.013.

Boretti, A., & Rosa, L. (2019) Reassessing the projections of the World Water Development Report. *NPJ Clean Water* 2, 15. https://doi.org/10.1038/s41545-019-0039-9

Bryman, A. (2008) *Social Research Methods*. Fifth. Oxford Uninversity Press.

Charmaz, K. (2006) *Constructing Grounded Theory*. 2006. SAGE Publications Ltd. https://uk.sagepub.com/en-gb/eur/constructing-grounded-theory/book235960 [Accessed: 5 October 2020].

Cook, C. & Bakker, K. (2012) Water security: Debating an emerging paradigm. *Global Environmental Change* 22 (1), 94–102. https://doi.org/10.1016/j.gloenvcha.2011. 10.011.

Given, L. (2008) In Vivo Coding. In: *The SAGE Encyclopedia of Qualitative Research Methods*. Thousand Oaks, CA, SAGE Publications, Inc. https://doi.org/10.4135/9781412963909.n240.

Global Water Partnership (2000) *Towards-water-security.-a-framework-for-action.-executive-summary-gwp-2000.pdf*. 2000. https://www.gwp.org/globalassets/global/toolbox/references/towards-water-security.-a-framework-for-action.-executive-summary-gwp-2000.pdf [Accessed: 29 November 2018].

Gunda, T., Hess, D., Hornberger, G.M. & Worland, S. (2019) Water security in practice: The quantity-quality-society nexus. *Water Security* 6, 100022. https://doi.org/10.1016/j.wasec.2018.100022.

Haque, M.M., Rahman, A. & Samali, B. (2016) Evaluation of climate change impacts on rainwater harvesting. *Journal of Cleaner Production* 137, 60–69. https://doi.org/10.1016/j.jclepro.2016.07.038.

Hoekstra, A.Y., Buurman, J. & Ginkel, K.C.H. van (2018) Urban water security: A review. *Environmental Research Letters* 13 (5), 053002. https://doi.org/10.1088/1748-9326/aaba52.

Jepson, W., Budds, J., Eichelberger, L., Harris, L., Norman, E., O'Reilly, K., Pearson, A., Shah, S., Shinn, J., Staddon, C., Stoler, J., Wutich, A. & Young, S. (2017) Advancing human capabilities for water security: A relational approach. *Water Security* 1, 46–52. https://doi.org/10.1016/j.wasec.2017.07.001.

Kahinda, J.M., Taigbenua, A.E. & Borotob, R.J. (2010) Domestic Rainwater Harvesting as an Adaptation Measure to Climate Change in South Africa. 2010. https://www.sciencedirect.com/science/article/pii/S1474706510001336 [Accessed: 11 July 2018].

Kisakye, V. & Van der Bruggen, B. (2018) Effects of climate change on water savings and water security from rainwater harvesting systems. *Resources, Conservation and Recycling* 138, 49–63. https://doi.org/10.1016/j.resconrec.2018.07.009.

Liaw, C.-H. & Tsai, Y.-L. (2004) Optimum storage volume of rooftop rain water harvesting systems for domestic use. *Journal of the American Water Resources Association* 40 (4), 901–912. https://doi.org/10.1111/j.1752-1688.2004.tb01054.x.

Mishra, A., Adhikary, A.K. & Panda, S.N. (2009) Optimal size of auxiliary storage reservoir for rain water harvesting and better crop planning in a minor irrigation project. *Water Resources Management* 23 (2), 265–288. https://doi.org/10.1007/s11269-008-9274-4.

Mohammad, M.J., Sai Charan, G., Ravindranath, R., Reddy, Y.V. & Altaf, S.K. (2017) Design, construction and evaluation of rain water harvesting system for SBIT Engineering College, Khammam, Telangana. *International Journal of Civil Engineering and Technology* 8 (2), 274–281.

Musayev, S., Burgess, E. & Mellor, J. (2018) A global performance assessment of rainwater harvesting under climate change. *Resources, Conservation and Recycling* 132, 62–70. https://doi.org/10.1016/j.resconrec.2018.01.023.

O'Hanlon, F. (2021). *Rainwater Harvesting and Community Water Security in south-west Uganda*. [Apollo – University of Cambridge Repository]. https://doi.org/10.17863/CAM.73570

Pacey, A. & Cullis, A. (1986) Rain water harvesting: the collection of rainfall and runoff in rural areas. https://onlinelibrary.wiley.com/doi/abs/10.1002/pad.4230080111.

Qadir, M., Sharma, B.R., Bruggeman, A., Choukr-Allah, R. & Karajeh, F. (2007) Non-conventional water resources and opportunities for water augmentation to achieve food security in water scarce countries. *Agricultural Water Management* 87 (1), 2–22. https:doi.org/10.1016/j.agwat.2006.03.018.

Rahman, A. (2017) Recent advances in modelling and implementation of rainwater harvesting systems towards sustainable development. *Water (Switzerland)* 9 (12), 959. https://doi.org/10.3390/w9120959.

SSWM (2018) *Decentralised Supply.* 2018. https://sswm.info/water-nutrient-cycle/water-distribution/hardwares/water-network-distribution/decentralised-supply [Accessed: 24 March 2020].

Staddon, C., Rogers, J., Warriner, C., Ward, S. & Powell, W. (2018) Why doesn't every family practice rainwater harvesting? Factors that affect the decision to adopt rainwater harvesting as a household water security strategy in central Uganda. *Water International* 43 (8), 1114–1135. https://doi.org/10.1080/02508060.2018.1535417.

Thomas, T.H. (2000) *DTU Publications.* 2000. https://www.eng.warwick.ac.uk/dtu2/pubs/rwh.html [Accessed: 22 November 2018].

UNICEF (2020) *Water.* 2020. https://www.unicef.org/wash/3942_4456.html [Accessed: 24 March 2020].

Vörösmarty, C.J., McIntyre, P.B., Gessner, M.O., Dudgeon, D., Prusevich, A., Green, P., Glidden, S., Bunn, S.E., Sullivan, C.A., Liermann, C.R. & Davies, P.M. (2010) Global threats to human water security and river biodiversity. *Nature* 467 (7315), 555–561. https://doi.org/10.1038/nature09440.

Water.org (2018) *Uganda's Water Crisis – Water in Uganda 2018.* 2018. Water.org. https://water.org/our-impact/uganda/ [Accessed: 31 July 2018].

WHO (2012) WHO | What is the Minimum Quantity of Water Needed? 2012. WHO. https://www.who.int/water_sanitation_health/emergencies/qa/emergencies_qa5/en/ [Accessed: 19 December 2018].

Zhang, S., Zhang, J., Jing, X., Wang, Y., Wang, Y. & Yue, T. (2018) Water saving efficiency and reliability of rainwater harvesting systems in the context of climate change. *Journal of Cleaner Production* 196, 1341–1355. https://doi.org/10.1016/j.jclepro.2018.06.133.

Appendix

STAKEHOLDER INTERVIEW GUIDE (WATER SECURITY FRAMEWORK)

Interview objectives

1. To identify, according to practitioners, the <u>key attributes of a water secure community.</u>
2. To explore how water practitioners implementing decentralised water provisioning strategies/technologies <u>understand</u> and <u>operationalise</u> the term water security.
3. To identify how community-level water security is currently assessed by water practitioners.

Interview Structure: Introduction

- Background on interviewee organisation including core areas of focus (WASH, children, vulnerable populations, protection, medical access etc.).
- Experience of interviewee (WASH specialist, M&E specialist, management, country coordinator etc.).

Water security knowledge and understanding:

- Have you come across the term 'water security'? If so, where?
- Could you explain what you believe water security refers to?
- Does your organisation have a specific interpretation of water security?
- In your opinion, how does water security differ from other terms used to describe water access?
- Can you describe some characteristics that make a population 'water secure'?

Current methods of assessment of WASH projects:

- Talk me through how WASH projects are currently assessed at your organisation.
- What metrics are used to measure success?
- What does a successful WASH project look like in your opinion?
- What are your organisation's core goals when implementing a water-provisioning project?
- Are the same assessment methods used for every project that your organisation carries out?

The role of water security in project assessment and implementation:

- Do you think current methods of assessment of the 'success' of water-provisioning interventions are adequate? If so, why? If not, why not?
- Do these methods help to identify the relationship between water and socio-economic development? Is this a relationship your organisation considers when implementing WASH projects?
- When implementing WASH projects how much of a priority is equitable access? How do you measure equitable access to a water source?
- When implementing WASH projects, how much of a priority is sustainability? Can you define what is meant by the sustainability of a WASH project?
- Do you ever consider whether a water-provisioning intervention can help beneficiaries to manage climate risk? Can you give examples of climate risks your beneficiaries face?

Decentralised water interventions:

- Given our discussion above, how do you think decentralised water services could contribute to water security for populations in low-resource settings?
- Do you view rainwater harvesting as a climate adaptation strategy? If so, why? If not, why not?
- What are some barriers to the implementation & adoption of rainwater harvesting for rural and urban communities in sub-Saharan Africa?

12 Rainwater Harvesting for Water Security and Gender Equity

Kemi Adeyeye and Aisha Bello-Dambatta

12.1 INTRODUCTION

Equitable access to water and sanitation is affected by climate and environmental change, gender, income and livelihood inequalities, infrastructure inequalities, population growth and urbanisation (Adeyeye et al. 2020). According to the World Health Organization, progress towards Sustainable Development Goals (SDG) 6: clean water and sanitation for all would need to be four times faster to avoid 1.6 billion people still lacking drinking water at home by 2030; five times faster in all rural areas globally to eliminate inequalities; ten times faster in the least developed countries; 23 times faster in fragile contexts; and 25 times faster in urban areas (WHO 2021). Women are at greater risk of climate impact and food and water insecurity than men in over two-thirds of nations, despite having inadequate representation, participation, and involvement in most water institutions (Herbert 2022). This situation has implications for gender equality and empowerment as women and girls are disproportionately affected by the lack of access to basic water, sanitation, and hygiene facilities due to periods of increased vulnerability, such as during menstruation and reproduction (Kayser et al. 2019). Women and girls also play more significant roles relative to men in water, sanitation, and hygiene activities, including in agriculture and domestic labour. Thus, the WHO defined four research priorities to address both SDG 5 (gender equality) and 6 (water and sanitation) (Kayser et al. 2019; Sommer et al. 2015; Geere et al. 2018; Ray 2007):

1. Research into women's water-fetching responsibility and time-use burden is required. This includes studies into the implications for health and economic well-being and the impact of unpaid domestic work on leisure, learning and other income-generating activities.
2. Studies are needed to better understand sexual harassment, gender-based violence and psycho-social stress due to unsafe water and sanitation facilities. Research is also needed to correlate the type, location and distance to facilities and the types and nature of gender-based harassment and violence.
3. Research into gender-based factors due to water and sanitation marginality is needed. Not being able to meet the increased need for hydration, hygiene and sanitation for women, especially during menstruation, pregnancy,

DOI: 10.1201/9781032638102-20

post-natal and caregiving, could have health, education, and psychosocial stress on women. It also impacts their ability to be productive citizens. Therefore, research disaggregated by demography, life stage and physiology, including the impact of physical work and exercise, is needed to better understand gender-based needs.

4. Research into women's social and political empowerment, including through participation in water, sanitation and hygiene decision-making and governance. Most studies (e.g. WHO 2021; Wallace and Cole 2020) document the role of women and girls in domestic water activities. However, gaps in the measures and extent of women's roles in securing, managing, and making decisions about water resources remain. Studies are needed to better understand how to increase participation and its impact.

This study aligns with points 1 and 4. The aim is to investigate water security and gender equity gaps through the lens of rainwater harvesting (RWH), using a water initiative in the south-eastern part of Madagascar as a case study. A review of place-based research by Gerlak et al. (2018) found relatively fewer studies on water security in South America and Sub-Saharan Africa. In Sub-Saharan Africa, studies addressed water security mostly at the community scale, with primary concerns of quantity, quality, and water for agriculture. The feasibility of RWH in Africa has received less attention, with studies often focusing on the design, simulation, and reliability aspects (Nandi and Gonela 2022). Significantly, the review by Nandi and Gonela (2022) found a low concentration of studies (about 10%) addressed human and societal factors and feasibility constraints such as gender, education, attitudes and social norms, income and livelihood, affordability, and pricing.

Therefore, this chapter presents a nuanced account of the interrelated socio-economic factors and the role of RWH in addressing them. The research questions are:

1. What is the socio-technical impact of RWH as a mechanism to both address water insecurity and improve livelihoods for women?
2. Does direct action to empower women through training and gainful employment improve their socio-economic status and their ability to influence community participation in water action and governance?

12.2 BACKGROUND AND RATIONALE

The first aspect of this study centres on issues of water security. Water security refers to the availability of an acceptable quantity and quality of water for health, livelihoods, ecosystems, and production, coupled with an acceptable level of water-related risks to people, environments, and economics (Grey and Sadoff 2007). It is also the capacity of a population to safeguard sustainable access to reliable, equitable and safe quantities of acceptable quality water (Shrestha and Aihara 2018; Gerlak et al. 2018):

- To sustain livelihoods, human well-being, and socio-economic development.
- To ensure protection against water-borne pollution and water-related disasters.
- To preserve ecosystems in a climate of peace and political stability.

Water security is connected to and dependent on social, economic, and environmental processes such as hydrologic, geographic, economic, environmental, social, political, legal and financial and can be measured at local, national, regional, and global scales (Mishra et al. 2021; Moumen et al. 2019). The challenges of water security extend beyond single-issue indicators such as hydrologic water stress to include governance, infrastructure and economic actors, institutions, macroeconomic structure, capacity development and resilience (Grey and Sadoff 2007; Shrestha and Aihara 2018). Thus, water security measures must address welfare, equity and water-related risks (Hoekstra, Buurman and van Ginkel 2018, in Adams et al. 2020). The social dimension of water security implies equitable access to water resources and services, including through robust policies and legal frameworks at all levels, the provision of robust and resilient infrastructure, and building the resilience of people and communities to cope with extreme water events (Van Beek and Arriens 2014). Thus, with the water security framework, SDG 6 recognises the links between water for life and livelihoods as part of a global nexus that connects climate, human, inter- and intra-national, food and energy security (Beck and Villarroel Walker 2013).

Secondly, community action through partnerships is explored as a means to improve gender participation in water action and governance. This aspect includes the range of political, technological, social, economic, and administrative systems necessary to develop, manage and deliver water services at different levels of society (*after* Rogers and Hall 2003). Cleaver and Hamada (2010) justify this holistic theoretical approach in their framework for gendered water governance, which also encompasses the gender-based analysis of water interactions. This framework, originally defined by Franks and Cleaver (2007), emphasises the focus on good water governance on resources, mechanisms and outcomes (Cleaver and Hamada 2010):

> "Resources include material resources (the natural environment, human labour and capacities, economic resources, and so on), and non-material resources, such as institutions, social structures, and systems of rights and entitlements. Resources are drawn upon in different ways by people – that is, 'actors' (individuals, groups, the state) who need access to water. The arrangements in which they do this are specific to their context. These ways of organising access to water are the 'mechanisms' of water governance. Mechanisms (sometimes referred to as 'instruments') include formal and informal institutions (such as committees, and collective labour groups), tariffs and fees, arrangements for queues, rotations, and technology. These mechanisms channel water access depending on gender and other social identities. For example, a tariff is not gender neutral because it requires the ability to command cash. Outcomes refers to the water, which is accessed, but also refers to other effects related to this: the impact on people's livelihoods, the impact on their ability to give voice to their needs and concerns, and to the capacity for collective action and community development. Importantly, different mechanisms also have various implications for the ecosystem".

Community partnerships with non-governmental organisations, private-sector agents, public utilities, and donors; technological innovations; and community autonomy are important for financial sustainability, asset security, and service improvements (Adams et al. 2020). Therefore, through the lens of a social enterprise in a case study area, this study also explores the efficacy of a socio-technical approach to both tackle water insecurity and gender participation and governance.

Thus, the third focus of this study is to explore the socio-technical impact of RWH as a mechanism to both address water insecurity and improve gender equity. RWH is the collection, storage, distribution, and reuse of rainwater for domestic, non-domestic and industrial purposes. Rainwater is considered 'clean' for non-potable use but would typically require filtration or treatment for potable use. RWH is considered sustainable because it provides simple and direct access to water with minimal infrastructure costs, provides storm water control by decreasing saturation and runoff, provides backup for emergencies such as peak water demand and fire, and most importantly, improves food security, reduces water-borne diseases, decreases distances for water access and creates jobs if included in wide-spread community development projects (de Sá Silva et al. 2022; Quinn et al. 2021). It is beneficial in places where the spatial distribution of precipitation is highly influenced by geography and climate dynamics, which in turn lead to areas with high water cyclicity, hydrological extremes of flooding and droughts (Moumen et al. 2019); all of which contribute to water scarcity in varying ways.

In this work, RWH is explored as one of the means to address water security; whereby scarcity is based on the idea that resources (means) are limited, whereas wants and desires (ends) are not (Beck and Villarroel Walker 2013). By addressing the three dimensions of water governance: resource, mechanism, and outcomes. Water insecurity is generally considered a technological problem centred on addressing water quantity and quality problems, and typically studies on RWH would focus on these factors. However, emerging paradigms now focus more on alternative tools such as the recycling and reuse of water, demand management, integrated flood/drought management, promoting blue/green infrastructure, increasing community and stakeholder participation, effective governance and a multidisciplinary approach to achieve water security (Misha et al. 2021; Gareau and Crow 2006). This expands the scope to study RWH in a gender- and place-based manner.

12.3 THE CONTEXT

Madagascar is the fourth-largest island in the world. In mid-2023, the population was estimated at 30,571,093 (Worldometer 2023) and an estimated 40% lived in urban areas. Madagascar is ranked about a third from last in access to safe water and fourth from last in access to sanitation (USAID 2022). The hydroclimate and geography of Madagascar means that it is prone to cyclones and floods to the east and extreme aridity and water stress to the west, all exacerbated by climate change. Storm events, heavy rainfall, and runoff (Figure 12.1) are common in some areas, often resulting in water turbidity and contamination. In recent years, southern Madagascar has experienced recurrent droughts due to population growth, increasingly intense land use, annual cyclones, and erratic rainfall, among other factors, resulting in repeated emergency responses to address food and water insecurity (USAID 2022).

The Malagasy population is very young: 50% are under 18, 43% under 15 and 15% under 5. Depending on the source, the average household size ranges between 4.5 and 6.5. Most men and women aged 15–49 are either married or in union (63% and 58%, respectively); single women and men account for 24% and 38% of the total,

(a)

(b)

FIGURE 12.1 Post-rainfall flooding observed in some areas (Photos taken in April 2022).

respectively. About 13% of children under 18 live with neither of their biological parents, 16% live only with their mothers and 4% with their fathers (UNICEF 2019).

Madagascar remains a high-priority country for the SDGs, notably SDG6. The SDG 6 metrics include water quality, distance to safe water supply, impact on domestic roles, etc. Time for fetching water is also used to define a basic water service, yet these data are rarely disaggregated by gender at the global level (WHO 2021). According to UNICEF (2019), over half of the Malagasy population (57%) has limited or no access to an improved water source. Four in five people do not have access to basic sanitation services, and open defecation still occurs in some areas. Only one person in four has a handwashing facility with soap and water. Four out of five people drink water contaminated with *E. coli* and faecal matter. 83% of the population resides in rural areas, of which only about 35% have improved water supply and 9% have improved sanitation. In urban areas, only 18% have access to safely managed sanitation facilities and 58% still use unimproved toilets.

For Madagascar, there is a dearth of up-to-date socio-technical water research. Boone et al. (2011) undertook a non-nationally representative study that surveyed households in the rural and urban areas of the six provinces in Madagascar. They found that in rural areas, 44% of households collect water from a stream, pond, river, or lake, compared to 8% in urban areas. In rural areas, they found on average that wells were the closest available source (660 m), followed by surface water sources (827 m), public taps (1,137 m), and private taps (1,799 m). Compared to urban areas, where distances were wells (230m), private taps (236 m), public taps (426 m) and surface water sources (1,303 m), respectively. The average years of education of the household head were 4.7 in rural areas and 7.5 in urban areas. The household head is female in 17% of rural households and 22% of urban households. Rural households on average were found to be considerably less wealthy than their urban counterparts: the mean of the asset index is −0.30 for the rural sample compared with 1.11 for the urban sample, a difference of about 1.5 standard deviations. They also found that, on average, adult women spend the most time in water collection, followed by girls, boys, and then adult men. Thus, women and girls will see the greatest benefits from interventions that substantially reduce the amount of time it takes to collect water (Boone et al. 2011).

12.4 METHODS

The study employed a socio-technical, place-based research approach. The research methods were interviews, workshops and ethnography. The resulting qualitative data was analysed abductively (Okoli 2023). Abduction enables the development of new theories from qualitative data underpinned by existing theories and concepts. It is different because it does not contain an explanation in itself (induction), nor does it constitute a new case of an already known general rule (deduction), but represents a combination of both (Vila-Henninger et al. 2022). The context and participants using purposive sampling. Purposive sampling is used where there is a need to gain detailed knowledge about a specific phenomenon rather than make statistical inferences and/or where the population is very small and specific (Obilor 2023). The participant views and responses were analysed to define categories of meanings that provide new understandings within existing theories (Jacob and Buijs 2011). Themes are typically used to derive a more implicit and abstract level, which requires interpretation, whilst categories refer to the explicit content of the text, often as a simple description of the participants' views and accounts (Vaismoradi et al. 2016).

The research design for the project is shown in Table 12.1.

TABLE 12.1
Research Design

Categories	Ethnography	Interviews (Semi-structured)	Workshop
Problem identification	X		
Risks, opportunities, and challenges	X	X	X
Current practices and impact	X	X	
Future change		X	X

12.4.1 CASE STUDY

This study was situated in the south-eastern part of Madagascar. Researchers from the UK partnered with a social enterprise in this region of Madagascar to investigate the use of technological innovation such as RWH to improve access to water and sanitation in the area as well as address gender and livelihood needs. The social enterprise's aim is to provide clean water for everyone in Madagascar. They work to maximise the potential of water resources in nature, i.e., rainwater, to improve water access in urban, peri-urban and rural areas in the south-eastern parts of Madagascar. As of 2023, they had installed 73 RWH systems in communities, hospitals, markets, and schools and delivered over 13 million litres of water directly to over 26,000 people. Their core strategy is to fund the installation of RWH infrastructure, then to train and use water agents to manage and maintain the water and work at the kiosks – a small timber structure where water is dispensed and sold to raise revenue (Figure 12.2). As part of their commitment to SDG 5 and 6, the social enterprise only recruits women for this role. There were 26 water agents when this study was conducted in 2022; however, this has grown to 46 by 2023.

12.5 SAMPLING AND DATA COLLECTION

Ethnographic studies were conducted in key zones of operation by the social enterprise (Figure 12.3, at the time of study) as well as in neighbourhoods and schools where other types of WASH provisions exist (Figure 12.4). A one-day multi-stakeholder workshop was also held, which was well attended by community groups, NGOs, academics from Madagascar and other countries, and regional policymakers.

This chapter focuses on and presents findings from semi-structured interviews with the water agents in the case study area. The study population is shown in Table 12.2, and details of the sample are in Table 12.3.

The age profile of the water agents was 23–40 years old, and they had been water agents for one month to two years. Only one was single and had no children. The majority were single parents with 2–3 children, and where disclosed, some had their first child as young as 15–17 years old. Contextually, most households are led by men. However, it is worth noting that multi-generational households were prevalent

FIGURE 12.2 (a) Rainwater harvesting installation at a school, with a nearby water kiosk and agent. (Photo taken in April 2022). (b) Another example of a rainwater dispensing kiosk used by water agents (faces obscured). (Photo taken in April 2022).

in the case study area, as observed during the ethnographic visits and interviews. As a place-based study, the impact of the RWH systems was explored in urban, peri-urban, and rural areas. Therefore, five of the water agents interviewed were based in urban areas, one in peri-urban areas and four in rural areas.

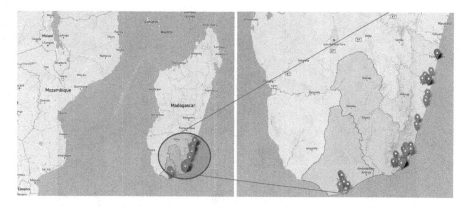

FIGURE 12.3 Rainwater harvesting installations at the time of this study.

The interviews were conducted within one week in April 2022. The primary target group were the water agents, and except for four that were interviewed at the office of the social enterprise, the others were interviewed at the water kiosks. Although they served as gatekeepers, neither the director nor the administrator (their line manager) were present when they were being interviewed. A trained interpreter was used to translate questions and responses from Malagasy to English. The interviews were also audio recorded and corroborated with notes taken by two researchers. The interview and resulting data were managed in line with ethically approved protocols. Where necessary, procedural statements by the water agents were corroborated with the data derived from interviews with the social enterprise director and project administrator.

12.6 FINDINGS

12.6.1 Rainwater Harvesting for Water Supply

According to the water agents, the actions of NGOs, private entities and social enterprises have improved access to water in their communities. Six of the ten interviewees source their water from the nearest RWH system. The others use independent providers primarily because they are 'closer to home' or where water is cheaper, or at no cost (corroborates the findings by Boone et al. 2011). For one: *I live next to (another water provider). My mother runs the kiosk, so we get free water".* For another: *"… I previously collected water from a well built by the community a while ago. A NGO added cement and made it safer. Now we use RWH from the kiosk".* Without these sources, the women stated that their closest water source is more than a 20–30 minutes' walk to the river or lake. *"Everyone in the town goes to this [provider name], or if there is no water, a nearby mountain stream. Or we go to someone's well but you can't guarantee the water. Sometimes people go with bike, cars etc. but if not, you walk or tuk-tuk if you can afford it." If no water, we go to collect water at a lake, and it is contaminated with mud during the drought, but we collect it anyway and use it. We sometimes also go to a nearby town… 20 mins walk away".*

FIGURE 12.4 (a) Child using a well (faces obscured). (Photo taken in April 2022). (b) Outdoor toilets for sanitation with a mural dissuading public defecation (Photo taken in April 2022).

Another said, *"We walk 20 mins when water is scarce. The river is closer but poor quality. RWH is a good option when there is rain".*

Due to the lack of robust water infrastructure, access to rainwater and travel distances were strongly linked to their water security. Most of the independently

TABLE 12.2
Study Population

Organisations	Stakeholder	Sample	Population	Percentage
Social enterprise (water)	Director	1	1	100
	Administrator	1	4	25
	Water Agents	11*	26	42

Ten reported due to limitations in the ability to communicate with one participant.

provided water sources are from wells, RWH systems, or, in extreme cases, water trucks. One lady said she travelled less distance when water was scarce because she knew someone with a well. But: *"Both the [RWH] tank and well runs dry during the drought. So, we have to walk to the mountain. We wake up early to do the 30 mins walk, and again in the afternoon. So, care is always taken not to waste water."* Conversely, most of the water agents currently live near or less than 5 minutes-walk from the RWH installation or similar improved water source.

At the time of the study, the cost of the harvested waster was 50 Ariary (AR) (circa 0.01 USD) per 20-L container and 100 AR (circa 0.02 USD) for 3×20-L containers. Therefore, the use of water (to drink or wash) is informed by both affordability and water quality. The interviewees said: *"We use different types of water for different activities e.g., 'tap' water for cooking and drinking, other sources for washing"*, and *"I use [social enterprise] water for drinking and cooking food, and bathing but when it is washing clothes etc., I go to the river* [WA08]". [WA05] stated that *"Water from river etc. is red. So some use [the RWH water] to drink and wash white clothes"*. Also, *"...there is a cost to [name of social enterprise] water so people only use it when the free well runs dry"*.

On average, the women said that they needed to make multiple trips to collect water. WA01, for example, said that she needs to make four trips per week with 20 L containers (100 L per week) to meet the needs of her household; another stated that they also made multiple trips to meet the 200 L per week water needs of her household. Most of the women say that they or their children are responsible for collecting and transporting water for the household. One went as far as to state that her children sometimes miss school to help her collect water.

Water adequacy is an important consideration for SDG 6, as it is important that the water quality is fit for purpose. The participants recognised the risks of poor water quality to their health: *"No clean water means a lot of problems. We use a dirty puddle of water or lake. We wash in it and get sickness from the water. There was never clean water even if you look for it. With this poor water, you get a sore on the skin which becomes cancerous, there is no healthcare as well"* [WA09].

It was interesting to find that the agents rarely treated water even for drinking, for instance by boiling before use: *"We do not treat water either from RWH or the river"* [WA03]. Instead, they use different sources of water for different applications within the home: *"RWH water is better for drinking etc. [name of another provider] water*

TABLE 12.3
Sample of Water Agents

Water Agent (WA)	Age	Education	Marital Status	Children	Home Status	How Long as Agent	Location	Work Tenure
01	36	Secondary school (finished)	Single parent	Girl – 10 Boy – 16	Lives with children	12 months	Urban	Part time
02	23	2 years of Secondary school.	Single parent	Girl - 8 Boy - 5	Lives with mother + 6 other siblings	Since Nov. 2021	Urban	Part time
03	39	1 year of secondary school	Single parent	Boy – 19 Girl – 16 Boy – 12 Boy - 9	Lives with children.	3 months, Jan 2022	Peri-urban	Part time
04	33	Last year of primary school (5 yrs)	Single parent	Boy – 13 Girl – 9	Lives with children	1 month (March 2022)	Rural.	Part time
05	27	1 year of secondary school	Married	Boy – 9 Girl – 6 Girl – 3	Lives with husband and children	2 years	Rural	Part time
06	28	Middle school	Married	Girl – 11 Girl – 9 Girl – 4	Lives with husband and children	9 months	Rural	Part time
07	26	Last year of high school	Single	No children	Lives with parents	9 months	Urban	Part time
08	29	Baccalaureate	Single parent	1 Child	Lives alone. Child lives with the parents	12 months	Rural	Part time
09	39	2 years of secondary school.	Single parent	1 Child. Plus her sister's (deceased) children	Lives with children	15 months	Urban	Part time
10	40	2 years of secondary school.	Single parent	1 Child	Lives with parents	2 years	Urban	Part time

is not so good" [WA04]. All the agents stated that the untreated RWH water quality was much better than that from other providers. The participants mentioned that there were 'little sediments and slime' at the bottom of the RWH container compared to other sources: *"Water from the well is not very clean. Cooking and drinking, I use 'tap' water [referring to the RWH water]. The one from the well, you can see it is unclean from the character of the water. So I wash first with the well water and finish with the tap water...* [WA09]". *"Other water source being 20 mins walk away, [Using RWH] saves time and effort. There is perception that this water is cleaner. Does not leave slime at the bottom of the bucket and it has better taste"* [WA01].

Confidence in the RWH water source also came from the fact that the water agents know the source and manage the infrastructure: *"I trust the quality of water provided by [social enterprise]. There is another kiosk in town but never tested it. I boil pump water before drinking but I don't treat the [name of social enterprise] water. I am the one that looks after the system, and I can tell others in the community too* [WA07]".

The water agents believed that the RWH installations have made a lot of difference in their communities as it helps to address the challenges of poor water quality and access: *"The community has not run out of water since the [social enterprise] system (in three months). Before then, water access was always in doubt. If available e.g. river, it is not clean due to pollutants and run off. But we have to use it anyway"*.

Most of the women were also aware of the changes in the environment and climate: *"Water is scarce during the dry season – July – September. Two years ago there was drought. We had to travel 20 mins to get water. With climate change, water shortages are getting worse. RWH is seasonal, what happens when it does not rain?"* Another said: *"With climate change, water shortages are getting worse"*.

12.6.2 RWH AND GENDER

The social enterprise in this case made the decision to only employ female water agents to manage and maintain their RWH system and kiosks. The social enterprise is a spin-off from an NGO that builds schools, especially in rural areas. Therefore, several of the RWH installations are located close to schools. The parent NGO works with the social enterprise to facilitate access and logistical challenges associated with construction work in remote, off-grid places. This social enterprise is also the first to employ women to construct the rainwater tanks. For instance, the system managed by WA04 was built entirely by a female construction team. It sits in a school situated in a remote area and was therefore built in two phases since getting materials and equipment to the site was very difficult. The RWH system also took two days to build, with the team living on site until it was completed. The researchers experienced the same difficulties when they visited the school; vehicles had to be abandoned, and the final stage of the journey involved wading through water.

The approach to recruiting the agents was disparate. Some heard of the role through word of mouth and made direct applications. Others were nominated by the *Chef de Fokontany* (The chief. The Fokontany is the smallest administrative unit in Madagascar, comprising one or several villages) or the headteachers where the RWH systems were located within local schools. Others came through their links with influential relatives. Some responses to the question of access to this form of employment were as follows:

- *Word of mouth from a friend who works at the social enterprise. I then applied and was successful.*
- *Word of mouth and the community leader offered me the job.*
- *I heard of [name of social enterprise] water agent job and applied.*
- *I spoke with the head of school and the chef fokontany, in [name of town], there was no one in charge of keeping track of the water and money in that location.*
- *The head of school in [the town] and my grandad is the president of the parents of school kids. That was how I heard of (name of RWH enterprise)*

It was observed that maintaining a delicate balance with the social or political figures within the communities when appointing agents engendered buy-in and other benefits for the social enterprise. However, this disparate, often subjective approach resulted in agents with different performative qualities (mostly positive) but also different abilities, e.g., numeric, written and communication skills (education background in Table 12.3). All agents are, however, given basic training, including how to record water meter readings, financial reporting and monitoring/maintaining the water systems.

For the women, running the RWH systems offered a different job type and prospects compared to the norm. Most said that they would have had to be resourceful in earning a living or take on low pay, or short-term jobs. For instance, *"I was running a kiosk where people can wash their hands during covid, after that I got this job"* [WA09]. It also provided an important source of livelihood for these women, who prior to this had to find other means such as washing clothes for others, working in a shop on low-paying, petty trading, etc. *"I was washing clothes. Income was poorer. Working in a shop. No sale, no pay. Without this job, I will be a fruit seller, sell bread to school kids, sell other items"* [WA01]. The water agent job was highly preferred because it offered status as well as a regular, guaranteed income: *"I needed the job because of the regular monthly pay. But I had to prove myself first"*. The job also came with other benefits, such as the social enterprise paying for their children's education. Although better paid, being a water agent remains a part-time job, so the women still do other jobs such as petty trading, subsistence farming and husbandry, crafts and weaving to complement their income. Some excerpts from the interviews were:

- *I raise chickens, subsistence farming [WA02].*
- *Subsistence farming – I grow rice but is not very good because of the wrong conditions; animal husbandry – chicks, zebu and pigs. Zebus are herded and they get water on the way. It takes a lot of water to keep pigs, as well as washing, cooking [WA03].*
- *I keep pigs, grow cassava, do weaving which I sell for additional income [WA04].*
- *Weaving. I was a weaver before this job. I sell the products in Mahatalaky market. I keep chickens, grows cassava [WA05].*
- *I was a petty trader, selling food like soup, small biscuits etc. [WA06].*

Madagascar still has traditional views about the role and position of women in society. For most of the water agents, their jobs helped to improve how they were perceived within their communities. Most were single parents, with the associated societal perceptions and challenges. Therefore, they felt empowered because they could independently support their families through a reliable source of income. Some said that they had savings for the first time: *"Since I have this job, I feel more empowered. I can save money to repair the house. But some people are happy for me, others jealous"* [WA02]. They also felt more respected:

> *"People give me more respect, we are more clean, have food, not the same as before. Big difference in the role of women, If you earn money, you can support the household, contribute. Otherwise, this is considered the role of men"* [WA05].
>
> *"I am considered more important and more lucky because I got the job. I have better livelihood; I can help my household. Role in [name of RWH enterprise] offers more position and communication opportunity because I am an agent. I am able to advice the community about the importance of the quality of water, better sanitation etc"* [WA06].

Others valued the flexibility of the part-time role: *"Happy because I am hardworking, and this job allows me to do other things as well as running the kiosks"* [WA03].

All the agents considered their training an invaluable aspect of their development: *"I was trained about the systems by [social enterprise] before the job. I feel like I have a voice in the organisation. I want to have more responsibility and grow in the organisation. I want to grow and make more money"* [WA05]. Others are looking forward: *"I would like to have communication training. I would like to do more in the organisation/ community. I believe that I can have more responsibility and earn more money..."* [WA06].

They also felt like they had a voice and could make contributions within the organisation and in society: *"I have a voice, they listen to the water agents e.g. if there is a problem, need new materials etc. [name of RWH enterprise] is very responsive too."* [WA06]. *"I want training where I can convince people in the bush on how to get good water. Good hygiene, training on people skills, and engagement for water agents. They think that the (unsafe) water is fine because it has been drunk for generations..."* [WA08].

12.6.3 WATER GOVERNANCE AND PARTICIPATION

The lack of structured and formalised water governance was observed. There were often no dedicated water champions, or water governance was part of other roles fulfilled by leaders in the communities: *"Only one person in the commune is in charge of water issues. No committees or discussions. People report issues to this person, and he then contacts the responsible person to resolve it."* Even though water security is recognised as an important issue, the agents mentioned that there were minimal forums or events where water resilience or action was the main agenda. There was only one exception: *"There was a meeting about water a couple of months ago. We were talking about how to keep the system and the water clean and safe for everyone. That is the only meeting we've had. There is also an association that teaches us how to keep the water system sustainable too. It would be good if we can have more meetings like this as it is good to educate the people and keep them informed. Most people here are women."* The water agents mentioned that information sharing tends to happen informally: *"People share information about the best places to collect water, nearest site, cost etc. No one to complain to, so people find ways to solve their problem".* In some areas, the kiosks are the hub for water discussions and debate. The water agent thus became an important water person in the community: *"There is no water meeting, but the community works together. Sometimes customers come to my house out of hours to ask me to open the kiosk. I also have a good relationship with the head of school".*

12.7 DISCUSSION

The findings in this chapter focus on the issues of gender-based water security, access, and participation. The findings are presented using the nexus of resources, mechanisms, and outcomes (Cleaver and Hamada 2010).

In the study area, it was found that water as a resource was variable due to natural effects, the lack of reliable infrastructure to deliver wholesome water and the challenges of distance and cost. Furthermore, women (especially those in single-parent households) experience water-related changes through the impact on their family's health and wellbeing, as well as their ability to maintain reliable access to education or sustain livelihoods. The RWH systems deployed by the social enterprise tackle these water issues, thus, to some extent, ameliorating the associated social-economic effects of water insecurity. With RWH, water access and the time and distances travelled to obtain water were found to be improved. In addition, the women agents had gainful employment with guaranteed salaries (whether it rains or not) and other benefits, including school fee subsidies for their children. Therefore, in addition to being more productive members of society, they can be independent and plan household finances rather than trying to survive from day to day. Therefore, RWH in this case has proven to deliver tangible outcomes for water security and gender equity.

However, some limitations were observed. Cost remained a primary issue for some households in accessing better-quality water. There was a charge for the water from RWH systems, so most households, including those of the water agents, often augment their water needs from free sources such as wells, rivers, or lakes. They often use water without further treatment, so they try to moderate the use of different sources of water for different needs and functions, e.g., rainwater harvested for potable uses, and other sources, e.g., for laundry. Due to this, RWH has minimal impact if distances are still being travelled to other sources or less quality water is being used due to cost. However, the extent to which this is a prevalent issue requires further study. Despite the disparate approach to recruiting the agents, most of them have proved to be effective in their roles. However, the high turnover rate for agents is a challenge for the social enterprise, including due to illiteracy. Recruiting and changing water agents are politically and socially sensitive, especially where agents are nominated by key social or political figures or, in some cases, are married to or related to them. Thus, one recommendation would be to utilise a more objective approach based on person and job specifications to recruit water agents.

Another observation, corroborated by the interview data, was that the opportunities for equitable participation in water governance beyond this localised level were limited. Despite its best efforts, the social enterprise's ability to improve this situation remained limited. This issue was explored through interviews with policymakers, and at the workshop, but the findings are beyond the scope of this chapter.

This chapter nevertheless contributes to the growing body of work on gender and water, with particular focus on access, quality, and women as vulnerable water users and/or women and water in rural communities (de Silva et al. 2018; Herbert 2022). The findings demonstrate that women's participation in meaningful water actions can improve their stewardship and entrepreneurial range, thus exerting tacit influence in actions and decision-making. Nevertheless, there remains a lagging gap in studies on wider gender participation in water governance (Ozano et al. 2022, cited in Herbert 2022) compared to other aspects of WASH. This is, however, changing, as reflected by the improving attitudes towards water agents. For instance, one of the communes visited during the study was led by the first female *Chef de Fokontany* in the region. Therefore, it is anticipated that better gender representation will emerge as norms and perceptions change. However, central and regional policymakers need to take the impetus to speed up the rate of change in water and other sectors. Future academic studies should also contribute by addressing the research gap in this area.

12.8 CONCLUSION

The WHO defined four priority areas to achieve SDG 6 goals for water security and gender equity as: minimising the time and distance burden on women and girls for fetching water; addressing sexual harassment, gender-based violence and psychosocial stress due to unsafe water and sanitation facilities; reducing water marginality so that women can be active, productive members of society; and improving social and political participation of women in water governance. This chapter detailed how a social enterprise is using RWH a socio-technical approach to deliver these priorities whilst empowering women and changing societal norms. Two research questions were defined at the onset of the study: First, to investigate the socio-technical benefit of RWH to address water access and security challenges whilst addressing gender equity and livelihood issues. Second, to determine whether this socio-technical RWH approach delivered a meaningful path to improving women's participation in water governance.

It was found that the approach employed by the social enterprise, which combined water and sanitation provision with women's empowerment, was highly successful. The women water agents, often single parents, gained both water and financial security. Their social acceptance and status in their communities also improved, and, in most cases, they became the source of information and advice on water issues in those communities. However, it was also observed that cost remained a barrier to access to quality water in these communities. People still made longer trips for less-wholesome water because it was free. Water provision was also found to be highly reliant on non-governmental agents. Although water policies exist, governance structures and processes, especially at the grassroots level, are less clear. Further, the path to improving gender, especially women's participation in water governance, was less clear. Therefore, a more joined-up effort between all governmental and non-governmental agents would be needed to maximise the impact of RWH in this and similar contexts and meet the widening water needs.

ACKNOWLEDGEMENTS

This project was funded by the Royal Academy of Engineering (RAENG) Frontiers Champions Tranche 2: FC-2122-2-48. The authors wish to thank the participants of this study, including the contributors that were outside the scope of this paper, the Malagasy people for their welcome and hospitality, and most importantly, Harry Chaplin and his colleagues in Madagascar.

REFERENCES

Adeyeye, K., Gibberd, J. and Chakwizira, J., 2020. Water marginality in rural and peri-urban communities. *Journal of Cleaner Production*, *273*, p.122594.

Adams, E.A., Zulu, L. and Ouellette-Kray, Q., 2020. Community water governance for urban water security in the Global South: Status, lessons, and prospects. *Wiley Interdisciplinary Reviews: Water*, *7*(5), p.e1466.

Asian Water Development Outlook (AWDO) 2016. Strengthening Water Security in Asia and the Pacific. Available online: https://www.adb.org/publications/asian-water-development-outlook-2016 (accessed on 01 November 2023).

Beck, M. B., & Villarroel Walker, R. 2013. On water security, sustainability, and the water-food-energy-climate nexus. *Frontiers of Environmental Science & Engineering*, *7*, 626–639.

Boone, C., Glick, P., & Sahn, D. E. 2011. Household water supply choice and time allocated to water collection: Evidence from Madagascar. *Journal of Development Studies*, *4712*, 1826–1850.

Cleaver, F., & Hamada, K. 2010. 'Good' water governance and gender equity: a troubled relationship. *Gender & Development*, *18(1)*, 27–41.

de Sá Silva, A. C. R., Bimbato, A. M., Balestieri, J. A. P., & Vilanova, M. R. N. 2022. Exploring environmental, economic and social aspects of rainwater harvesting systems: A review. *Sustainable Cities and Society*, *76*, 103475.

de Silva, L., Veilleux, J. C., & Neal, M. J. 2018. The role of women in transboundary water dispute resolution. *Water Security Across the Gender Divide*, 211–230.

Dominique, K., Trabacchi, C., Chijiutomi, C., Tshabalala, Z., Joshi, D., Udalagama, U., Nicol, A. and the GCRF Water Security Hub 2022. Groundwater: Making the invisible visible. *FCDO Briefing Pack on Water Governance, Finance and Climate Change, K4D Briefing Note. Institute of Development Studies.* Online at: https://opendocs.ids.ac.uk/opendocs/handle/20.500.12413/17243. Accessed: 03 November 2023.

Franks, T., & Cleaver, F. 2007. Water governance and poverty: a framework for analysis. *Progress in Development Studies*, *74*, 291–306.

Gareau, B.J.; Crow, B. Ken Conca. 2006. Governing Water: Contentious Transnational Politics and Global Institution Building; *International Environmental Agreements*; MIT Press: Cambridge, MA, USA, p. 466.

Geere, J. A. L., Cortobius, M., Geere, J. H., Hammer, C. C., & Hunter, P. R. 2018. Is water carriage associated with the water carrier's health? A systematic review of quantitative and qualitative evidence. BMJ Global Health, *33*, e000764.

Gerlak, A. K., House-Peters, L., Varady, R. G., Albrecht, T., Zúñiga-Terán, A., de Grenade, R. R., ... & Scott, C. A. 2018. Water security: A review of place-based research. *Environmental Science & Policy*, *82*, 79–89.

Grey, D., & Sadoff, C. W. 2007. Sink or swim? Water security for growth and development. Water Policy, *96*, 545–571.

Herbert, S. 2022. *Women's Meaningful Participation in Water Security.* GSDRC & K4D Helpdesk Report, University of Birmingham 20 May 2022. Online at: https://opendocs.ids.ac.uk/opendocs/bitstream/handle/20.500.12413/17467/1159_Women_meaningful_participation_in_water_security.pdf?sequence=5&isAllowed=y. Accessed: 03 November 2023.

Hoekstra, A. Y., Buurman, J., & van Ginkel, K. C. 2018. Urban water security: A review. *Environmental Research Letters*, 135, 053002.

Jacobs, M. H., & Buijs, A. E. 2011. Understanding stakeholders' attitudes toward water management interventions: Role of place meanings. Water Resources Research, *47*.

Kayser, G. L., Rao, N., Jose, R., & Raj, A. 2019. Water, sanitation and hygiene: measuring gender equality and empowerment. *Bulletin of the World Health Organization*, 976, 438.

Mishra, B. K., Kumar, P., Saraswat, C., Chakraborty, S., & Gautam, A. 2021. Water security in a changing environment: Concept, challenges and solutions. *Water*, *13*(4), 490.

Moumen, Z., El Idrissi, N. E. A., Tvaronavičienė, M., & Lahrach, A. 2019. Water security and sustainable development. *Insights into Regional Development*, 14, 301–317.

Nandi, S., & Gonela, V. 2022. Rainwater harvesting for domestic use: A systematic review and outlook from the utility policy and management perspectives. *Utilities Policy*, *77*, 101383.

Obilor, E. I. 2023. Convenience and purposive sampling techniques: Are they the same. *International Journal of Innovative Social & Science Education Research*, *111*, 1–7.

Okoli, C. 2023. Inductive, abductive and deductive theorising. *International Journal of Management Concepts and Philosophy*, *163*, 302–316.

Ozano, K.; Roby, A.; MacDonald, A.; Upton, K.; Hepworth, N.; Gorman, C.; Matthews, J.H.; Quinn, R., Rougé, C., & Stovin, V. 2021. Quantifying the performance of dual-use rainwater harvesting systems. *Water Research X*, *10*, 100081.

Ray, I. 2007. Women, water, and development. *Annual Review of Environment and Resources*, *32*, 421–449.

Rogers, P., & Hall, A. W. 2003. Effective Water Governance Vol. 7. Stockholm: Global water partnership. TEC background papers No.7. Online at: https://dlc.dlib.indiana.edu/dlc/bitstream/handle/10535/4995/TEC+7.pdf?sequence=1. Accessed: 03 November 2023.

Shrestha, S., Aihara, Y., Bhattarai, A.P. 2018. Development of an objective water security index and assessment of its association with quality of life in urban areas of developing countries. *SSM Population Health*, 6, 276–285.

Sommer, M., Ferron, S., Cavill, S., & House, S. 2015. Violence, gender and WASH: spurring action on a complex, under-documented and sensitive topic. *Environment and Urbanization*, 271, 105–116.

UNICEF. 2019. Madagascar Multiple Indicator Cluster Survey, 2018. Executive summary of survey results. Online at: https://www.unicef.org/madagascar/media/3121/file/UNICEF%20Madagascar%20Executive%20Summary%20MICS%20ENG.pdf. Published August 2019, Accessed 03 October 2022.

USAID 2022, Magagascar Global Water Strategy 2022–2027, Online at: https://www.globalwaters.org/sites/default/files/madagascar_hpc_plan.pdf. Accessed: 03 November 2023.

Van Beek, E., & Arriens, W. L. 2014. Water Security: Putting the Concept into Practice. p. 55. Stockholm: Global Water Partnership.

Vaismoradi, M., Jones, J., Turunen, H., & Snelgrove, S. 2016. Theme Development in Qualitative Content Analysis and Thematic Analysis. *J Nurs Educ Pract.* 6: 100–10. https://nordopen.nord.no/nord-xmlui/bitstream/handle/11250/2386408/Vaismoradi.pdf?seque

Vila-Henninger, L., Dupuy, C., Van Ingelgom, V., Caprioli, M., Teuber, F., Pennetreau, D., ... & Le Gall, C. 2022. Abductive coding: theory building and qualitative re analysis. *Sociological Methods & Research*, 18, e0279651.

Wallace, T., & Coles, A. 2020. Water, gender and development: An introduction. In *Gender, water and development* pp. 1–20. Routledge.
Worldometer 2023, Madagascar population. Online at: https://www.worldometers.info/world-population/madagascar-population/. Accessed: 03 November 2023.
World Health Organisation (WHO). 2021. Progress on household drinking water, sanitation and hygiene 2000–2020: Five years into the SDGs, Situation report. Online at https://www.who.int/publications/i/item/9789240030848 Published: 13 September 2021; Accessed: 29 September 2022.

Case Study 6
Madagascar
Social Enterprise as a Financially Sustainable and Replicable Model for Long Term Functionality

Harry Chaplin

SUMMARY

Tatirano Social Enterprise is an organisation empowering women by improving access to clean drinking water in southern Madagascar. A large part of Tatirano's activities is rainwater harvesting at schools, hospitals and in communities. Tatirano aims to combat the trend of water infrastructure falling into disrepair over time as a result of inadequate management. In order to ensure functionality and proper monitoring, a local lady is typically employed in the position of a Tatirano agent in order to monitor and manage the water system and sell water to the local community. To pay for this ongoing management and repair of water infrastructure, Tatirano also sells a high-end water product from its treatment centre – bottled water without the single-use plastics. Tatirano uses the profit made from high-end water sales to pay for overhead costs and ongoing repairs and monitoring of all water infrastructure under its portfolio. This social business model, where 100% of the profits are ploughed into the ongoing functionality of water systems at schools and in communities, is not only proving appropriate for scaling in Madagascar but is replicable for all contexts where there is a bottled water demand and a lack of affordable and improved community water access.

PROBLEM STATEMENT

Water infrastructure in poor countries around the world without the means to manage, monitor and repair effectively quickly falls into disrepair. This leads to millions of US dollars being wasted as infrastructure is used for a matter of months and then left to rot. Establishing a financially sustainable water business alongside

DOI: 10.1201/9781032638102-21

social outputs has begun to prove an effective and replicable model for the long-term functionality of water infrastructure. Consistent and long-term functionality of water infrastructure then has the opportunity to impact communities over time on health benefits and time savings as initially intended.

Tatirano water systems currently provide clean water for drinking and handwashing in 28 schools, 2 hospitals and at 38 community kiosks. A further 16 schools are planned or under construction. Water at the taps throughout school sites is free and unlimited for students and teachers. Eighteen of the school sites are hybrid school-community kiosk systems, where there is also a gravity-led pipeline to a community kiosk on the edge of the school property. These systems are managed by a local lady in the position of Tatirano Agent – she manages the rainwater infrastructure, resets the first flush systems and sells water in the community kiosks. At the time of writing, these community kiosks provide water for approximately 2,000 households.

Using regional school data, manually collected user data and average household census data, over 11,000 students and 10,000 community members have access to clean water at Tatirano systems when there is rainfall. Rainfall on the east coast of Madagascar is regular and significant (>100 mm per month); however, climate change is making weather patterns more unpredictable and extreme, leaving smaller water systems empty for 2–3 months per year.

The south and southeast of Madagascar follow strong traditional and cultural norms. The majority of water collectors in high-rainfall areas where Tatirano works in rainwater harvesting are women and girls. Therefore, an increase in access to clean water closer to home benefits women especially, and hence the premier objective of Tatirano is to empower women *via* access to clean water.

ACTIONS TAKEN

Tatirano installs variations of rainwater harvesting systems centralised around the construction of a highly efficient water tank design – the Calabash. Pioneered in Guinea-Bissau by Paul Akkerman and colleagues in 2010, the Calabash is a ferrocement tank used in rural and urban rainwater harvesting projects in many countries across Africa and Latin America. Tatirano has adapted the design to be locally appropriate to the context and needs, choosing to install mostly 10,000 and 20,000 L Calabash tanks instead of the popular 5,000 L variety.

These above-ground storage tanks are coupled with large roof areas at schools or marketplaces to distribute water by gravity to various water points for free usage by the schools and paid for by community users. A very simple first flush system is installed to protect the quality of the rainfall on the roof as it enters the tank, and microbial analyses have suggested that this can be sufficient to achieve WHO guidelines.

Local ladies are employed to manage the water points and to feed data into a public statistics platform called Statirano. All data for systems is publicly available, making the organisation's activities transparent to partners and making it easy for the operations and maintenance teams in Tatirano to monitor and repair systems quickly after they have a problem.

A small water treatment centre closes the loop and hopes to provide financial sustainability for the social enterprise by treating rainwater and selling it for profit to cross-subsidise expenses for ongoing maintenance.

RESULTS

Tatirano seeks maximum functionality in the water systems that it installs and repairs. Tatirano installs very simple rainwater harvesting infrastructure that is conducive to longevity and has limited maintenance requirements. However, there is always the need for some level of ongoing maintenance in a water system, and once the external funding for capital expenditure, i.e., installation costs, has finished, the major challenge of ongoing functionality is the sustainability of finances for labour, materials and transport to carry out the repairs.

Tatirano has found that selling water in a high-rainfall area at community kiosks does not yield significant revenue. This is due to the fact that the price is kept low by a high supply of surface and other alternative water sources. Whereas in low-rainfall areas the value of water is greater, Tatirano has found that a higher price can be charged, and revenues received are high enough for a sustainable business model in ongoing management.

As a result of this learning, in high-rainfall areas where Tatirano deploys rainwater harvesting as the sole water infrastructure, the treatment and sale of high-end water are important to fill the financial gap between functionality costs and kiosk revenue.

The success of Tatirano Social Enterprise should be measured on its ability to provide a high uptime level – the period that a water system is functional and providing clean water – and its ability to continue to pay for this through the money-making aspects of its model.

SUSTAINABLE DEVELOPMENT GOALS AND IMPACT

Tatirano's social business model described in this case study contributes to two main Sustainable Development Goals: SDG 5: Gender equality and SDG 6: Clean water and sanitation.

At the time of writing, January 2024, 14,498,130 L of clean water had been supplied in schools and throughout communities in the southeast of Madagascar. Approximately 30,000 people use water from Tatirano systems every week. Tatirano calculates this from school student register data and by counting the number of users and multiplying that figure by the average household size for the area.

Tatirano trained and now runs a mixed construction team, made up of inspiring young women in senior mason and management roles that encourage all women and girls everywhere to dream and act outside of traditional expectations and norms.

Ultimately, and not yet completely measured or proven for Tatirano, increased access to clean water has profound and lasting effects on the productivity and well-being of a community. Tatirano's work, providing clean water closer to homes and schools, means that fewer school and work days are missed as a result of illnesses associated with dirty water.

Women, being central to water collection and management and the management of ill family members, clearly gain significantly more time for a range of community-lifting activities such as spending more time with children or earning more money.

NEXT STEPS

Tatirano intends to continue to improve the installation and durability of the water systems in the first instance and the response time of repairs to ensure the maximum uptime possible of the water systems. This will improve the current portfolio of systems and their functionality.

In order to reach more people, Tatirano has three strategies going forward:

1. Tatirano will increase the rate of new installations to six systems per month;
2. Following the successful replication of the Thai Jar, Tatirano will explore innovative marketing and financing to ensure that as many households as possible can access the Madagascar Jars of 1,000 and 2,000L. The target price is USD 50 for a portable 2,000-L ferrocement jar – approximately 10% of the price of plastic alternatives in Madagascar;
3. Tatirano will begin to run awareness campaigns and practical workshops with communities about basic and important rainwater harvesting principles (gutter positioning, tank sizing, first flush, maintenance for water quality, etc.)

In order to further Tatirano's financial sustainability, more investment will be made into the collection and storage capacity of Tatirano's water treatment centre. More water treatment centres will be installed in nearby towns to enter new geographical markets and raise sales.

ADVICE FOR OTHERS LOOKING TO DO SOMETHING SIMILAR

Remember why you are doing it. Hold true to your passion and your beliefs. Tatirano is made up of the community and grounds itself in the belief that development opportunities arise from the presence of convenient, clean water. Belief and confidence in local skills and knowledge are keys to success. Secondly, make mistakes and errors and continue to make things slightly imperfect until the model improves. Tatirano is far from having achieved a fully financially sustainable model but is headed in the right direction, and whilst Tatirano is still reliant on some donor funding, one day the model will be strong enough to maintain itself.

Index

Note: **Bold** page numbers refer to tables and *italic* page numbers refer to figures.

Printed in the United States
by Baker & Taylor Publisher Services